D0271658

Two week

loan

Genes and
Resistance to Disease

Springer

*Berlin
Heidelberg
New York
Barcelona
Hong Kong
London
Milan
Paris
Singapore
Tokyo*

V. Boulyjenkov K. Berg Y. Christen (Eds.)

Genes and Resistance to Disease

With 28 Figures and 23 Tables

 Springer

Boulyjenkov, V., Dr.
Human Genetics
Department of Noncommunicable Disease Prevention
World Health Organization
1211 Geneva 27
Switzerland

Berg, K., Prof. Dr.
Institute of Medical Genetics
University of Oslo and Department of Medical Genetics
Ullevål University Hospital
Oslo
Norway

Christen, Y., Ph. D.
Fondation IPSEN
Pour la Recherche Thérapeutique
24, rue Erlanger
75781 Paris Cedex 16
France

ISBN 3-540-66724-5   Springer-Verlag Berlin Heidelberg New York

Library of Congress Cataloging-in-Publication Data
Genes and resistance to disease / V. Boulyjenkov, K. Berg, Y. Christen (eds.). p. ; cm. – Includes index. ISBN 3540667245 (hardcover : alk. paper) 1. Medical genetics-Congresses. 2. Genes-Congresses. 3. Natural immunity-Congresses. 4. Immunogenetics-Congresses. I. Boulyjenkov, V. (Victor), 1948- II. Berg, Kåre. III. Christen, Yves. IV. World Health Organization. V. Fondation IPSEN pour la recherche thérapeutique. [DNLM: 1. Genetics, Medical-Congresses. 2. Gene Therapy-Congresses. 3. Immunity, Natural-Congresses. QZ 50 G3218 2000]
RB155 .G3589 2000
616'.042–dc21

Springer Verlag is a company in the Bertelsmann Springer publishing group.

© Springer-Verlag Berlin Heidelberg 2000
Printed in Germany

Production: PROEDIT GmbH, 69126 Heidelberg, Germany
Cover design: design & production, 69121 Heidelberg, Germany
Typesetting: Mitterweger & Partner GmbH, Plankstadt
Printed on acid-free paper – SPIN: 10717926   27/3136wg – 5 4 3 2 1 0

Foreword

Since 1989, the World Health Organization (WHO) and the Ipsen Foundation have organized joint meetings, bringing together small groups of experts concerned about genetic issues in medicine. These meetings are not intended to lead to any official decisions, and they bind neither WHO nor the Ipsen Foundation. Rather, their objective is to taking a bearing on a medical domain with important implications for all. That is clearly the case for this meeting, because the notion of a protective gene is both innovative and abounding in possibilities for application.

The advances in human genetics that have occurred during the past 20 years have revolutionized our knowledge of the role played by inheritance in health and disease. It is clear that our DNA determines not only the emergence of catastrophic single-gene disorders, which affect millions of persons worldwide, but also interacts with environments to predispose individuals to cancer, allergy, hypertension, heart disease, diabetes, psychiatric disorders and even to some infectious diseases.

However, such progress often evokes fears of control of peoples' reproductive behavior. Scientific, medical and lay communities should ensure that information and technology will be used to preserve the dignity of the individuals and that adequate controls exist in countries to prevent abuses of genetic information and unacceptable practices.

Environmental factors and genetics, separately and together, have brought out many risk factors for diseases. This is, after all, the standard medical approach: it considers disease first and foremost. In theory, however, another approach is possible: identifying not risk factors but protective factors.

Overall, the study of longevity and the demonstration of genes favoring a long lifespan suggest that such protective systems exist. In recent years, the study of genetic polymorphisms has made clear that some alleles have beneficial effects. These discoveries can substantially improve our understanding of the interactions between genetics and the environment, between pathogenetic mechanisms and new treatments based upon these protective mechanisms, either by using the genes themselves in gene therapy or, more generally, by mimicking the underlying protective biochemical processes.

These issues are of great importance and were discussed extensively at a joint WHO/Fondation IPSEN meeting held in Venice, 7-8 February 1999, and devoted to the problems of possible genetic resistance to diseases. The papers included in this volume have been produced within the framework of a WHO/Fondation IPSEN meeting; they express the views of the individual authors rather than a consensus of the participants at the meeting. These views also do not represent the policies of the WHO or the Fondation IPSEN.

Victor Boulyjenkov   Yves Christen
                                                                      April 2000

Contents

Genetics of Survival
F. Schächter 1

The Human Genes that Limit AIDS
S. J. O'Brien, M. Dean, M. Smith, C. Winkler, G. W. Nelson, M. P. Martin,
and M. Carrington 9

Gene Protecting Against Age-Related Macular Degeneration
P. Amouyel 19

Genes Involved in Resistance to Carcinogenesis
C. R. Wolf. 27

Leptin and Neural Circuit Regulating Body Weight
J. M. Friedman 43

The Cholesteryl Ester Transfer Protein (CETP) Locus
and Protection Against Atherosclerosis
K. Berg 51

Does the Gene Encoding Apolipoprotein A-I Milano Protect the Heart?
C. R. Sirtori and L. Calabresi 67

Candidate Gene Polymorphisms in Cardiovascular Pathophysiology
F. Cambien 83

Protective Response of Endothelial Cells
M. P. Soares, C. Ferran, K. Sato, K. Takigami, J. Anrather, Y. Lin,
and F. H. Bach. 91

Genetic Factors in Malaria Resistance
L. Luzzatto 105

Genetic Basis of Resistance to Alzheimer's Disease and Related
Neurodegenerative Diseases
C. L. Masters and K. Beyreuther 121

Genes Affecting Cognitive and Emotional Functions
P. McGuffin 133

Gene Therapy: Promises, Problems and Prospects
I. M. Verma, L. Naldini, T. Kafri, H. Miyoshi, M. Takahashi, U. Blömer,
N. Somia, L. Wang and F. H. Gage 147

„Good Gene"/Bad Gene"
V. I. Ivanov 159

Gene protecting against cancers and Tumor Suppressor Genes
J. J. Mulvihill 169

Subject Index 179

List of Authors

Amouyel, P.
INSERM U508, Institut Pasteur de Lille, 1 rue Albert Calmette, 59019 Lille Cedex,
France

Anrather, J.
Immunobiology Research Center, Beth Israel Deaconess Medical Center, Harvard
Medical School, Boston, MA 02215, USA

Bach, F.H.
Immunobiology Research Center, Beth Israel Deaconess Medical Center, Harvard
Medical School, Boston, MA 02215, USA

Berg, K.
Institute of Medical Genetics, University of Oslo and Department of Medical
Genetics, Ullevål University Hospital, Oslo, Norway

Beyreuther, K.
Center for Molecular Biology, The University of Heidelberg, 69120 Heidelberg,
Germany

Blömer, U.
Laboratory of Genetics, The Salk Institute, 10010 North Torrey Pines Road,
La Jolla, CA 93037, USA

Calabresi, L.
Center E. Grossi Paoletti, Institute of Pharmacolgical Sciences, University of
Milano, Via Balzaretti 9, 20133 Milano, Italy

Cambien, F.
INSERM U 525, 17 rue du Fer à Moulin, 75005 Paris, France

Carrington, M.
Laboratory of Genomic Diversity, National Cancer Institute, Frederick Cancer
Research and Development Center, Building 560, Room 21–105, Frederick, MD
21702–1201, USA

Dean, M.
Laboratory of Genomic Diversity, National Cancer Institute, Frederick Cancer Research and Development Center, Building 560, Room 21–105, Frederick, MD 21702–1201, USA

Ferran, C.
Immunobiology Research Center, Beth Israel Deaconess Medical Center, Harvard Medical School, Boston, MA 02215, USA

Friedman, J.M.
Howard Hughes Medical Institute, The Rockefeller University, 1230 York Avenue, Box 305, New York, NY 10021, USA

Gage, F.H.
Laboratory of Genetics, The Salk Institute, 10010 North Torrey Pines Road, La Jolla, CA 93037, USA

Ivanov, V.I.
Research Center for Medical Genetics and the Russian State Medical University, 1, Moskvorechie str., 115478 Moscow, Russia

Kafri, T.
Laboratory of Genetics, The Salk Institute, 10010 North Torrey Pines Road, La Jolla, CA 93037, USA

Lin, Y.
Immunobiology Research Center, Beth Israel Deaconess Medical Center, Harvard Medical School, Boston, MA 02215, USA

Luzzatto, L.
Department of Human Genetics, Memorial Sloan Kettering Cancer Center, New York, NY 10021, USA

Martin, M.P.
Laboratory of Genomic Diversity, National Cancer Institute, Frederick Cancer Research and Development Center, Building 560, Room 21–105, Frederick, MD 21702–1201, USA

Masters, C.L.
Department of Pathology, The University of Melbourne, Parkville and the Mental Health Research Institute of Victoria, Victoria 3052, Australia

McGuffin, P.
Social, Genetic and Developmental Psychiatry Research Centre, Institute of Psychiatry, De Crespigny Park, 111 Denmark Hill,London SE5 8AF, UK

Miyoshi, H.
Laboratory of Genetics, The Salk Institute, 10010 North Torrey Pines Road, La Jolla, CA 93037, USA

Mulvihill, J. J.
University of Oklahoma Health Sciences Center, Oklahoma City, OK, USA

Naldini, L.
Laboratory of Genetics, The Salk Institute, 10010 North Torrey Pines Road, La Jolla, CA 93037, USA

Nelson, G.W.
Laboratory of Genomic Diversity, National Cancer Institute, Frederick Cancer Research and Development Center, Building 560, Room 21–105, Frederick, MD 21702–1201, USA

O'Brien, S. J.
Laboratory of Genomic Diversity, National Cancer Institute, Frederick Cancer Research and Development Center, Building 560, Room 21–105, Frederick, MD 21702–1201, USA

Sato, K.
Immunobiology Research Center, Beth Israel Deaconess Medical Center, Harvard Medical School, Boston, MA 02215, USA

Schächter, F.
CESTI – ISCM, Pole Universitaire Léonard de Vinci, 92916 Paris La Defense Cedex, France

Sirtori, C. R.
Center E. Grossi Paoletti, Institute of Pharmacolgical Sciences, University of Milano, Via Balzaretti 9, 20133 Milano, Italy

Smith, M.
Laboratory of Genomic Diversity, National Cancer Institute, Frederick Cancer Research and Development Center, Building 560, Room 21–105, Frederick, MD 21702–1201, USA

Soares, M. P.
Immunobiology Research Center, Beth Israel Deaconess Medical Center, Harvard Medical School, Boston, MA 02215, USA

Somia, N.
Laboratory of Genetics, The Salk Institute, 10010 North Torrey Pines Road, La Jolla, CA 93037, USA

Takahashi, M.
Laboratory of Genetics, The Salk Institute, 10010 North Torrey Pines Road, La Jolla, CA 93037, USA

Takigami, K.
Immunobiology Research Center, Beth Israel Deaconess Medical Center, Harvard Medical School, Boston, MA 02215, USA

Verma, I. M.
Laboratory of Genetics, The Salk Institute, 10010 North Torrey Pines Road, La Jolla, CA 93037, USA

Wang, L.
Laboratory of Genetics, The Salk Institute, 10010 North Torrey Pines Road, La Jolla, CA 93037, USA

Winkler, C.
Laboratory of Genomic Diversity, National Cancer Institute, Frederick Cancer Research and Development Center, Building 560, Room 21–105, Frederick, MD 21702–1201, USA

Wolf, C. R.
Imperial Cancer Research Fund Molecular Pharmacology Unit, University of Dundee Biomedical Research Centre, Ninewells Hospital and Medical School, Dundee DD1 9SY, UK

Genetics of Survival

F. Schächter

Summary

The fields of gerontology and genetics have merged, spawning novel lines of investigation and generating a wealth of new results in recent years. However, the lack of clarity and consistency in the basic definitions upon which the science of gerontology must rest has fostered a certain amount of enduring confusion. Among the unclear issues are the genetic components of life span and the distinction between "normal" and "pathological" aging. At a time of massive world population aging such issues have, beyond their scientific importance, a momentous social and economic impact. Here I propose a simple axiomatic framework, consisting of eight postulates and definitions, that clarifies the above-mentioned issues and reconciles disparate data in gerontology. Based on this framework, a new classification of genes involved in survival is proposed.

Introduction

After years of promoting the "genetics of longevity," I realized at length how misguided I had been! This is a terrible misnomer and I believe it has helped lead astray other scientists interested in human gerontology, for a simple reason: longevity is the outcome of a lifetime of manifold events, integrating all types of intrinsic and extrinsic influences on viability – or vulnerability, a term that conveys the downward trend of an aging organism. What makes it so misleading is that the apparent simplicity of a simple scalar value – longevity – hides the diversity of its contributing components. One might as well try to understand a complex three-dimensional landscape through its projection on a straight line. A statement closer to reality can be outlined in four stages: 1) mortality is age-dependent, 2) age-dependent mortality is partially under genetic control, 3) the adjective "partially" makes room for environmental contributions and 4) the genetic components of age-dependent mortality are themselves age-dependent. Survival, as a function of age – of which mortality is the first derivative with a minus sign – lends itself to the necessary distinctions. Therefrom derives a new framework calling for new questions:
- What are the major genetic components of age-dependent mortality?
- Are these components interrelated? In what ways?

V. Boulyjenkov, K. Berg, Y. Christen (Eds.)
Genes and Resistance to Diseases
© Springer-Verlag Berlin Heidelberg 2000

- Can a new classification of genes involved in aging be derived?
- What is to be learnt from the genetics of age-dependent mortality about aging processes, and vice-versa?
- What are the present frontiers of survival and what paths of exploration are opened?

In what follows, we start addressing these questions.

The Compensatory Adaptation (CA) Theory of Aging

The tenets of this theory have been described elsewhere (Schächter 1997, 1998). Here I shall present the theory in the condensed form of a set of postulates and definitions, then propose some predictions to be checked in future investigations, and finally offer one biological illustration.

Postulate 1: The genetic design of the organism is optimized for pre-reproductive fitness.

Postulate 2: The mere functioning of an organism brings about irreversible alterations in its molecular, cellular, and tissular components.

Postulate 3: The genome itself is the target of some irreversible and cumulative damage with chronological time, which entails a progressive loss or alteration of genetic information.

Postulate 4: In an effort to survive, the organism reorganizes its metabolism to compensate for its deficiencies.

Definition 1: Aging is the composite outcome of irreversible alterations and partially reversible compensations. It always corresponds to a decrease in viability, or an increased vulnerability.

Definition 2: Primary aging is the core component of aging resulting from irreversible alterations undergone by mere functioning.

Definition 3: Compensatory adaptation (CA) refers to reorganizations of metabolic pathways and coordinated functions consecutive to primary aging or to external stresses. It is the second component of aging.

Postulate 5: Both components of aging are under partial genetic control.

Now let us briefly comment upon these postulates and definitions.

Postulate 1: The genetic design is understandable in light of the whole life history, from conception through development and reproduction, with their respective time scales. This is Darwinian evolution in a nutshell. Implicit in this postulate is the dwindling evolutionary pressure on the post-reproductive period of an organism's life span and its consequences in terms of the evolutionary appearance of aging (Williams 1957).

Postulate 2: In fact, this postulate has a much wider generality as it applies to any physical system. Biological systems have the unique property of being able to repair and renew themselves to some extent. Their maintenance systems are not sufficient, however, to fend off cumulative damage, because of postulate 1. Therefore, the rate of alteration is the result of two opposite processes: damage and repair.

Postulate 3: This postulate is the basis of the old somatic mutation theory of aging. Although we know this to be true, information about the sequence and tissue specificity of such damage remains very scant. Age dependence also remains to be plotted out. The key question at this point is the extent to which genomic damage contributes to aging.

Postulate 4: Individual adaptation stands in clear distinction from adaptation in the evolutionary sense. The latter has been the molding force of the germ-line-transmitted genome; it contains the blueprint of every situation that was faced in the past, with solutions that were devised to cope and survive. All pathways in their intricate coordination can function perfectly in the internal milieu of a pre-reproductive organism with most of its genetic information intact. Individual adaptation, however, is a solution put together with defective components so that it is, by necessity, sub-optimal.

Definition 1: This is a biological definition of aging at the individual level. It ties up with the formal statistical definition at the population level in terms of an increase in mortality rates. From this definition stems the belief that aging is not synonymous with the passage of time, since some compensations may be reversed and some damage may even be repaired, thereby reducing the irreversible kernel of aging. This definition offers the advantage that it directly points to processes and mechanisms, opening the way for action.

Definition 2: It is noteworthy that primary aging is not synonymous with intrinsic or endogenous aging. Indeed, under the best environmental conditions that minimize damage from external sources, some molecular damage will still accumulate, due to free radical reactions, etc. This damage, resulting from basic combustion reactions, cannot be attributed exclusively to either external or internal sources that are nearly impossible to separate in an open system. Therefore, there is little sense in talking about endogenous aging. Intrinsic aging includes some of the consequences of primary aging under conditions of minimal external stress.

Definition 3: CA derives both from primary aging and from supplementary stresses. One might say that the CA component of aging contains (intrinsic minus primary) aging plus the lasting effects of other stresses. The latter covers pathological aging. The definitions could be summarized in the following equation:

Total Aging = Primary Aging + CA = Intrinsic Aging + Pathological Aging

Postulate 5: Both components of aging may display environmental influences and gene × environment interactions, although these are probably far more important for the second component.

Here are some predictions of the theory:

- Modulation of aging rate and therefore of maximum life span could be accomplished by modulating one of the two terms of the damage-repair equation;
- The common denominator of aging manifestations in a given species is the compound of primary aging and minimal resulting CA;
- Given the characteristic common patterns of aging, the distribution of damage in time and space may exhibit a much greater specificity than is currently thought;

– By identifying the targets of primary aging, it may become feasible to compensate directly for the incurred damage and therefore minimize the CA component of aging.

A biological illustration of CA is cellular senescence. Ever since the seminal work of Hayflick and Moorhead (Hayflick 1965), laboratories have been fascinated with the "Hayflick limit," researching the mechanisms behind the cells' limited division capacity. Telomere shortening is one of the more studied genomic alterations that correlates well with cellular senescence (Harley et al. 1990), in the sense that the doubling potential may be modulated to some extent by switching on or off the telomerase activity. Other genomic alterations, such as a reorganization of chromatin before entry into the senescent stage, have received less attention, although they are probably at least as important. Innumerable papers have fuelled the never-ending debate as to whether cellular senescence is a cause or a consequence of aging. The alternative to cellular senescence is the division of cells containing altered genetic information, leading to cancer and other clonal disorders (Sing and Reilly 1993). Pre-senescent cells may be forced to continue dividing until a tiny proportion of the culture will finally yield immortalized cells, while the remainder of the culture undergoes a massive death crisis. These data argue clearly in favor of the compensatory adaptive nature of cellular senescence as it occurs during aging. Indeed, senescent cells are associated with pathological rather than normal aging (Macieira-Coelho 1995). The progressive emergence of senescent cell populations is a common but not universal feature of aging, as seen, for example, in centenarians (Boucher et al. 1998).

A Novel Classification of Genes Involved in Survival

This classification is naturally derived from the CA theory of aging. It consists of separating the genetic contributions into the primary and CA components of aging.

1. Genes involved in primary aging

These genes modulate one term of the damage-repair equation. Intervening in the rate of damage are genes of the endocrine pathways regulating basal metabolism, genes that provide the first line of defense against reactive by-products of the metabolism, encompassing the common antioxidant enzymes. Intervening in maintenance and repair are DNA repair genes, which have been extensively characterized in several organisms, but also genes involved in protein and lipid repair and turnover.

In this category of genes, a change in activity would affect many other pathways, necessitating a coordinate regulation. Therefore, little genetic variation is expected. If a variant increased life span, it would do so in diverse environments. Because these genes depend on basic energy metabolism in their activity, putative longevity-extending variants could display antagonistic pleiotropy with fitness (Kirkwood and Rose 1991). Interactions are expected with the environment whenever it affects rates of damage or maintenance. Such is the case with nutritional intake or exposure to chemical pollutants or radiation. It is noteworthy that caloric restriction falls into that category by increasing the turnover rates of proteins and lipids, and maybe even of DNA.

2. Genes involved in CA

Many examples of CA can be envisioned at all levels of biological organization. They are obviously interrelated, which in itself is sufficient to account for the compensation effect on mortality and the negative correlations between the incidences of major age-related pathologies (Gavrilov and Gavrilova 1991). Among these, one may cite the increase in cardiac cell volume at the cellular level, the increase in left ventricular volume at the organ level, and the hypercoagulation phenotype found in centenarians (Schächter 1998).

Because these compensations occur at the expense of one another, a fair amount of genetic variation is expected. The effects should be largely dependent on the environment and individual life history. Antagonistic pleiotropy between survival at different periods of the life time may take place (Toupance et al. 1998).

To this day, the so-called human longevity genes have fallen into the second category. Therefore, the "paradoxes of longevity," whereby the risk status of certain alleles or genotypes in adulthood turns into a protective status at very old ages, should come as no surprise.

In the nematode *C. elegans,* however, genes of the first category have been found. It is greatly interesting that one of these genes bears homology to an insulin receptor gene and another to a mitochondrial component of the respiratory chain (Wood 1998). They appear to plug right into the energy metabolism.

The "Rate-of-Living" Theory of Aging Revisited

This is perhaps the most enduring theory of aging, formulated by Raymond Pearl in the early part of our century. Like the sea serpent, this theory re-emerges periodically under a different guise. It is based on the observation that life span is, by and large, inversely proportional to the basal metabolic rate.

I want to show that this theory is contained in the CA theory of aging herein presented. Primary aging is but one term in the equation of aging. It is already the result of two categories of processes: damage and alterations on the one hand; maintenance, replacement and repair on the other hand. Let us label these terms symbolically by:

DA for the global rate of damage and alteration processes,

MR for the global rate of maintenance, replacement and repair processes.

DA depends on the biochemical and biophysical parameters of a living system: temperature, pH, redox potential. DA has been estimated for various types of DNA damage (Lindahl 1993). There is a good case for it being proportional to the rate of oxygen consumption, because free radicals generated during respiration are the main culprits of biochemical damage. What remains to be assessed, though, is the extent to which other biochemical pollutants pervading our present environment and entering the body through the atmosphere and food – the fuel components – may increase this term, or, stated otherwise, what is the minimal value and the range of variation of *DA?* We know also that *DA* may be increased by exposure to certain radiations, and the possibility that it might be reduced by specific radiations cannot be discarded, although there are no data about this interesting hypothesis at present.

MR on the other hand is an evolved response to *DA* encoded in the genome. The whole machinery of *MR* can be defined in terms of networks of genes. There is no biochemical or biophysical reason *a priori* why *MR* should not be able to completely counter *DA,* except that it makes no evolutionary sense for any species to have developed such a system (Kirkwood and Rose 1991). Whereas *DA* is largely conserved across species, an endless variety of combinations exists to modulate *MR.* This explains why the inverse correlation between metabolic rate and life span holds only across related species. For example, flying birds have a much higher life span energy potential (total amount of energy available in a life time, equal to the product of the maximum life span potential and the specific metabolic rate) than mammals of similar size (Perez-Campo et al. 1998). This is associated with higher antioxidant activity, better mitochondrial coupling resulting in less leakage of free radicals, and greater genome stability.

Genome Dynamics on Two Time Scales

A fundamental polarity lies at the heart of evolution and aging: the balance between plasticity and stability of the genome. Evolution would not be possible without the appearance of new genetic diversity in the germ line at each generation. Yet such diversity includes the pool of genetic variants that are responsible for inherited diseases and the pool of slightly deleterious mutations that remain more covert but make up the genetic load (Kondrashov 1995). Providing an estimate of the deleterious mutation rate is an open and actively investigated problem (Eyre-Walker and Keightley 1998).

The progressive and unavoidable contamination of species' genomes by late-acting deleterious mutations is regarded as the major evolutionary mechanism accounting for the universal occurrence of aging across animal phyla (Partridge and Prowse 1994; Partridge and Barton 1993). The same basic phenomena are at work in the soma. They are summarized in the interplay between *DA* processes and *MR* systems. Sexual recombination may provide a means of eliminating deleterious mutations in bunches, thereby escaping the genetic decay that could threaten our mutation-prone race (Crow 1997, 1999). But sexual reproduction may also impose a limit to longevity, because genomes from male gametes that have undergone too many divisions bear a certain genetic load (Gavrilov et al. 1997). We may be led to discover hitherto unsuspected mechanisms of "averaging selection" that would prevent mutants of increased longevity from taking over, such as negative correlations between genetically determined early and late mortality. While these correlations have emerged as properties of a simple mathematical model (Toupance et al. 1998), their biological substrate is far from elucidated.

Finally, now that we know better that genetic components of survival should be segregated according to the different segments of the life time affected, much remains to be learned about these components from basic epidemiologic studies.

References

Boucher N, Dufeu-Duchesne T, Vicaut E, Farge D, Effros RB, Schächter F (1998) CD28 expression in T cell aging and human longevity. Exp Gerontol 33:267–282

Crow JF (1997) The high spontaneous mutation rate: is it a health risk? Proc Natl Acad Sci USA 94:8230–8386

Crow JF (1999) The odds of losing at the genetic roulette. Nature 397:293–294

Eyre-Walker A, Keightley PD (1998) High genomic deleterious mutation rates in hominids. Nature 397:344–347

Gavrilov LA, Gavrilova NS (1991) The biology of lifespan: a quantitative approach. Harwood Academic, Chur, Switzerland

Gavrilov LA, Gavrilova NS, Kroutko VN, Evdokushkina GN, Semyonova VG, Gavrilova AL, Lapshin EV, Kushnareva YE (1997) Mutation load and human longevity. Mutat Res 377:61–62

Harley CB, Futcher AB, Greider CW (1990) Telomeres shorten during ageing of human fibroblasts. Nature 345:458–460

Hayflick L (1965) The limited in vitro lifetime of human diploid cell strains. Exp Cell Res 37:614–636

Kirkwood TBL, Rose MR (1991) Evolution of senescence: late survival sacrified for reproduction. Phil Trans R Soc London B 332:15–24

Kondrashov AS (1995) Contamination of the genome by very slightly deleterious mutations: Why have we not died 100 times over? J Theoret Biol 175:583–594

Lindahl T (1993) Instability and decay of the primary Structure of DNA. Nature 362:709–715

Macieira-Coelho A (1995) The implications of the "Hayflick limit" for aging of the organism have been misunderstood by many gerontologists. Gerontology 41:94–97

Partridge L, Barton NH (1993) Optimality, mutation and the evolution of ageing. Nature 362:305–311

Patridge L, Prowse N (1994) Mutation, variation and the evolution of ageing. Curr Biol 4:430–432

Perez-Campo R, Lopez-Torres M, Cadenas S, Rojas C, Barja G (1998) The rate of free radical production as a determinant of the rate of aging: evidence from the comparative approach. J Comp Physiol B 168:149–158

Schächter F (1997) Genetics of aging. In: Robine JM, Vaupel JW, Jeune B, Allard M (eds) Longevity: to the limits and beyond. Springer, Heidelberg, pp 131–138

Schächter F (1998) Causes, effects and constraints in the genetics of human longevity. Am J Human Genet 62:1008–1014

Sing CF, Reilly SL (1993) Genetics of common diseases that aggregate, but do not segregate, in families. In: Sing CF, Hanis CL (eds) Genetics of cellular, individual, family, and population variability. Oxford University Press, Oxford, pp 140–161

Toupance B, Godelle B, Gouyon P-H, Schächter F (1998) A model for antagonistic pleiotropic gene action for mortality and advanced age. Am J Human Genet 62:1525–1534

Williams GC (1957) Pleiotropy, natural selection, and the evolution of senescence. Evolution 11:398–411

Wood WB (1998) Aging of C. elegans: mosaics and mechanisms. Cell 95:147–150

The Human Genes that Limit AIDS

S. J. O'Brien, M. Dean, M. Smith, C. Winkler, G. W. Nelson,
M. P. Martin and M. Carrington

Summary

The development of AIDS symptoms is a gradual process whereby a diverse population of HIV-1 genomes replicate in macrophages, monocytes, and T cells, challenging the immune system to its extreme. The cellular compartments and machinery that facilitate the process are human gene products that are punctuated by allelic variation that in some cases, determines the efficiency and kinetics of disease progression. Using molecular genetic typing of epidemiologic cohorts of HIV-1-infected study participants, we have searched for host genetic variants in genes whose products participate in HIV-1 replication. To date we have discovered attributable genetic influence on HIV-1 infection, disease progression, and AIDS sequelae involving coding and promoter regions of several human genes, namely *CCR5, CCR2, SDF1, HLA-A, -B, and -C*. This report will highlight the discovery, characterization and functioning of the multi-genic influences on the outcomes of HIV-1 infection and the influence of these variants on epidemiologic heterogeneity of the AIDS epidemic.

The scourge that we know as AIDS first became apparent in the late 1970s when a cluster of unusual cancers called Kaposi's sarcoma began to appear as a prelude to immune collapse among sexually active homosexual men in San Francisco, Los Angeles, New York and Washington. The immune deficiency disease was soon surmised and subsequently proven to result from infection by a newly emerged, T-lymphocyte-tropic retrovirus named HIV-1. In the past two decades HIV-1 has spread throughout the world, infecting up to 40 million people and resulting in at least eight million deaths, according to the latest reports of the World Health Organization (Bozzette et al. 1998; Mann and Tarantola 1998). Sub-Saharan Africa has seen the greatest impact, accounting for some 21 million infections, and mortality is rapidly growing in southeast Asia. The increasing death toll has stimulated a disease-related adjustment of world demographic projections downward for the first time since the Black Death of the thirteenth century (Warvick 1998).

Concern over the global impact of the epidemic has stimulated a surging research emphasis on the causes and on intervention strategies aimed at combating AIDS, but to date there remains no effective vaccine or therapeutic cure. Nonetheless, the field of AIDS research has witnessed a number of important and promising advances that have raised at least the hope that we may one day under-

V. Boulyjenkov, K. Berg, Y. Christen (Eds.)
Genes and Resistance to Diseases
© Springer-Verlag Berlin Heidelberg 2000

stand the process of AIDS pathogenesis and perhaps defend against it. Among the encouraging developments are the following:

First, the introduction of powerful combination anti-retroviral therapies that have reversed the accelerating population mortality for HIV-1-infected persons at least in Western countries (Bartlett and Moore 1998; Palella et al. 1998). Unfortunately, these drugs, however potent, do not completely eliminate the virus of infected patients as HIV-1 remains sequestered in tissue reservoirs that seem to avoid drug-mediated clearance of virus. Also, the cost of the therapy is too high to be available in the developing world where mortality, morbidity and viral spread continue unabated.

Second, the demonstration by cell biologists, virologists and x-ray crystallographers over the last three years that certain chemokine receptors serve as requisite HIV-1 co-receptors has opened a new avenue for intervention, raising the prospect of developing therapeutic compounds that attack the host cellular component, now seen as a necessary collaborator in the gradual and deliberate destruction of the infected patient's immune system (Cairns and D'Souza 1998; Cohen 1997; Littman 1998; McNicholl et al. 1997; O'Brien 1998).

Third, the demonstration that HIV-1 replicates at a remarkable rate, producing over one billion virions per day for up to 10 years before incapacitating the T-lymphocyte armada, gave compelling credence to the suspicion that the healthy immune system, particularly the HLA-restricted, cell-mediated immunity, was waging a fierce, albeit futile, defense against a highly adaptive predatory virus (Collins et al. 1998; Ho et al. 1995; Rosenberg et al. 1997; Wei et al. 1995).

The progression from HIV-1 infection to AIDS can best be described as a collaboration. The parties include the HIV viral population, proliferating and generating a genetic swarm at a dazzling speed. The second partner is the infected person, the host, providing a milieu of cellular havens plus all the machinery required for the virus to traverse its journey from a quiescent invader nestled in macrophages to a virulent predator decimating T cells that it infects along with others that it does not. Before we explain our approach to the discovery of genetic influences on the virus-host pathological process, it is useful to review a few points about the conventional wisdom concerning our own genomes.

The human genome consists of some 70,000–100,000 genes, ranging in length from a few hundred to several million nucleotides, arranged in linear arrays along 24 unique human chromosomes. Today a worldwide scientific endeavor called the Human Genome Project is most of the way to its 15-year goal: to produce a full length sequence of the 3.2 billion base pairs that comprise the human genome (Collins et al. 1997, 1998). The project, scheduled for completion in the coming year, has already uncovered considerable genetic variation among people and peoples. The project has revealed relatively common genetic variants every 500–1000 nucleotides (Chee et al. 1996; Lipshutz et al. 1999; Wang et al. 1998). This translates into an estimated 1.5–2.0 million genetic differences between any two people. Relating these differences to human phenotypes offers the alluring prospect for a genetic understanding of the basis for human distinctions in

Table 1. Genes that affect HIV-1 infection, AIDS progression and AIDS outcome

	Allele	Mode	Effect	Time	Citation
CCR5	Δ32	Recessive	Prevent infection	–	Dean et al. 1996
CCR5	Δ32	Dominant	Prevent lymphoma	Late	Dean et al. 1999
CCR5	Δ32	Dominant	Delay AIDS	Overall	Dean et al. 1996
CCR5P	P1	Recessive	Accelerate AIDS	Early	Martin et al. 1998
CCR2	64I	Dominant	Delay AIDS	Overall	Smith et al. 1997
SDF1	3'A	Recessive	Delay AIDS	Late	Winkler et al. 1998
HLA	A, B, C Homozygosity	Co-dominant	Accelerate AIDS	Overall	Carrington et al. 1999
HLA	B*35	Co-dominant	Accelerate AIDS	Overall	Carrington et al. 1999
HLA	C*04	Co-dominant	Accelerate AIDS	Overall	Carrington et al. 1999

appearance, talent, behavior, hereditary disease, inflammatory reaction and our immune response to infectious diseases including HIV.

Over the past few years, our very talented collaborators within and outside of the National Cancer Institute-USA have extended the critically important discoveries around chemokine cell biology, HIV co-receptors and cellullar immunity to identify human gene variants that directly influence the outcome of exposure to and infection with HIV-1 (Carrington et al. 1999; Dean et al. 1996; Martin et al. 1998; O'Brien and Dean 1997; Smith et al. 1997; Winkler et al. 1998). We have now identified seven different human loci (Table 1), originally candidate genes that were surmised to have a role in AIDS progression, by integrating seminal advances in four separate biological disciplines. These are: 1) the definition of the critical requirement for chemokine receptors to act as portals along with previously described CD4 molecules for HIV-1 entry into macrophages, monocytes and T lymphocytes; 2) the intricate and complex detailing of clinical conditions of thousands of HIV-1-exposed and -infected individuals using the best tools of AIDS epidemiology; 3) the application of the sophisticated biotechnology developed by the molecular genetic community for polymorphism discovery and high throughput genotyping and 4) the computational algorithms composed to utilize 50 years of population genetic theory to the task of associating common gene polymorphisms with disease outcomes. The synthesis of these approaches has revealed a number of genes that limit or accelerate AIDS in patients and shed important light on the mechanisms that are involved in this complex deadly disease.

In order to understand the role of the newly described AIDS restriction genes, it is important to first explain the view of AIDS pathogenesis and HIV-1 infection that was resolved within the past few years. A number of critical discoveries has led to these inferences and the details of their experimental approaches have been reviewed elsewhere (Chan and Kim 1998; D'Souza and Harden 1996; Littman 1998; McNicholl et al. 1997; O'Brien and Dean 1997). When patients initially become infected with HIV-1, they encounter primarily an M-tropic (for macrophage-tropic, also called R5) strain of HIV-1 that enters macrophages, monocytes and a subset of T lymphocytes which the virus invades by binding first to CD4 on the cell's surface. This causes a conformational change of the virus gp120 coat glycoprotein so that it now binds specifically to a chemokine co-receptor, CCR5, a process that facilitates membrane uptake of the virus itself. Some infected cells are not killed by the virus but rather become factories for HIV-1 production sequestered in part from effective clearance by HLA-restricted CD8+ cytotoxic T lymphocytes. Over one billion virions per day are produced and most, but not all, are dispatched by an active immune system challenged by the hordes of new virus particles. In the majority of patients, after many years of infection, a new variant of HIV appears, a T-tropic (also called T-cell or X4) virus (Berger et al. 1998; Connor and Ho 1994; Richman and Bozzette 1994; Schuitema-ker et al. 1992). This virus enters T cells using CD4 and a second chemokine receptor, CXCR4. The X4-tropic virus is more virulent and cytotoxic than the M-tropic strain and is more effective at killing T cells. The appearance of the X4-tropic virus generally precedes the decline of the CD4 T-lymphocyte count, a sign that the immune system is being irreversibly destroyed. The natural ligands of CCR5 are chemokines called RANTES, MIP1α and MIP1β, proteins which when bound to CCR5 effectively interfere with R5 HIV-1 infection. The ligand of CXCR4 is a powerful chemokine called stromal derived factor (SDF-1) that alter-natively will block entry of X4 virus but not R5-tropic virus. In all there are some 20 chemokine receptors described to date and about 60 chemokines (Premack and Schall 1996). There are several minor HIV co-receptors that allow certain strains of HIV to infect cells, but CCR5 and CXCR4 seem to be the principal co-receptors (Berger et al. 1998; Cairns and D'Souza 1998; D'Souza and Harden 1996).

The approach taken by our group and others was to assemble large cohorts of patients at risk for HIV-1 infection and to create B-cell lines from each as a renewable store of DNA for population genetic association analysis. In all, we have worked with some 20 different AIDS cohorts, including homosexual men, IV drug users and hemophiliac patients exposed by contaminated clotting factor supplies before the introduction of the HIV blood test in 1984 (Buchbinder et al. 1994; Detels et al. 1996; Goedert et al. 1985, 1989; Hilgartner et al. 1993; Kaslow et al. 1987; Phair et al. 1992; Vlahov et al. 1991; Lederman et al. 1995). In total, our laboratory has collected over 6,000 individuals who have formed the basis for our findings. We first screened DNA samples from a number of study partici-pants to detect common DNA polymorphisms within or near candidate genes such as the chemokine receptors and their chemokine ligands. After discovering a variant, we determined the frequency distribution of alleles and genotypes in different disease categories. A disease category is a subdivision of an AIDS cohort defined by clinical outcome, e.g., HIV infected vs. uninfected, or rapid progres-sors vs. slow progressors to AIDS after infection. We also performed epidemio-logic survival analyses where the rate of AIDS onset after HIV infection is com-pared between different genotypes. The approach was attempted for scores of candidate genes as well as for several hundred anonymous markers that were polymorphic in human populations. The results of these analyses have been published by our group (Table 1) and in most cases by other laboratories as well (Huang et al. 1996; Kostrikis et al. 1998; Lee et al. 1998; Liu et al. 1996; McDermott et al. 1998; Michael et al. 1997; Mummidi et al. 1998; Samson et al. 1996; Zimmerman et al. 1997). The details of these discoveries have been described elsewhere and will not be repeated here. Rather, we shall briefly summarize the status of these gene variants to date.

Early in 1996, several research groups, stimulated by the demonstration of the role of CCR5, CXCR5, CCR2 and other chemokine receptors in HIV-1 infection, began searching for genetic variants in the structural genes. Four groups almost simultaneously uncovered a common genetic variant in *CCR5,* a 32-base pair deletion within the coding region (Dean et al. 1996; Liu et al. 1996; Samson et al. 1996; Zimmerman et al. 1997). Because the deletion altered the reading frame for the transcript, it led to a truncated version of the receptor protein which never appeared on the cell surface of homozygotes for the deletion. When we tested our cohorts we did not find any *CCR5-Δ32/Δ32* homozygotes among over 1800 HIV-infected patients, although the frequency among uninfected individuals was 1–2 %. This discrepancy in genotype frequency could only have been interpreted as a reflection of the genetic resistance of these homozygous individuals to HIV-1 infection. In addition individuals who carried one normal and one mutant copy of the *CCR5* gene did become infected but they progressed to AIDS more slowly than those with two normal alleles, avoiding AIDS for two to four years longer on average. Thus, the *CCR5-Δ32* mutation seemed to confer a recessive genetic restriction to HIV-1 infection and a dominant (weak, but statistically significant) delay in AIDS onset. Within the next few years several laboratories affirmed this result in typings of over 16,000 individuals, virtually assuring the powerful genetic restraint on AIDS provided by carriers of the *CCR5-Δ32* mutation.

Using a similar approach we also described mutations in two other loci, *CCR2* and *SDF1,* whose alleles were also shown to slow AIDS onset among infected patients, although neither had a demonstrable effect on HIV-1 infection (Smith et al. 1997; Winkler et al. 1998). The functional basis of genetic restriction by *CCR2* and *SDF1* is not certain but the circumstantial data collected so far suggest that the *CCR2* allele may affect the quantity or availability of the primary receptors CCR5 or CXCR4 in the cellular infection targets of HIV-1-positive individuals (Lee et al. 1998). The *SDF1* mutation is a variant located in an evolutionarily conserved portion of the 3'-untranslated region of the *SDF1 β* gene transcript. The simplest but still unproven hypothesis is that the *SDF1* variant differentially interacts with a cellular DNA or RNA binding factor that alters the transcription or persistence of the *SDF1* mRNA, perhaps overproducing SDF-1β in some cellular compartment and preventing the spontaneous emergence of the late-stage X4-tropic HIV-1 types.

There are also a number of single nucleotide variants in the promoter region of the *CCR5* gene that have been shown to alter the rate of progression to AIDS as well (Kostrikis et al. 1998; Martin et al. 1998; McDermott et al. 1998; Mummidi et al. 1998). The most common haplotype, termed *CCR5P1,* seems to accelerate the development of AIDS among homozygotes infected with HIV-1. One of the single nucleotide variants defined by the *CCR5* promoter haplotype variant actually has been seen to have differential affinity for nuclear DNA binding factors, posing a potential mechanism for quantitative regulation of *CCR5* expression (Bream et al. 1999), already known to be a rate-limiting factor for both infection and AIDS progression. Finally, some recent studies using molecular-based typing of *HLA* alleles have demonstrated a powerful effect of locus heterozygosity on the rate of AIDS progression (Carrington et al. 1999). Thus, individuals who display locus heterozygosity for any combination of alleles at *HLA-A, -B* or *-C* delay AIDS for significantly longer periods than do patients who are homozygous for one or more of these loci. In addition, two alleles, *HLA B*35* and *Cw*04,* are significant risk factors for rapid progression to AIDS, whereas some 60 other *HLA* alleles appear to have little or no effect.

In sum, at least seven separate genes (Table 1) contribute in a slight but discernible way on various aspects of this devastating infectious disease. The attributable risk for these genes has been computed and together they account for between 30 and 50 % of the long-term survivors, those fortunate patients who avoid AIDS for 10 or more years after infection. This means that over 50 % of the heterogeneity in epidemiological outcome of HIV-1 exposure or infection is yet to be discovered. We speculate that other genes that have additional limiting steps on AIDS will be uncovered in the near future using this approach. The present results testify to the power of the well-described and documented AIDS cohorts (Buchbinder et al. 1994; Detels et al. 1996; Goedert et al. 1985, 1989; Hilgartner et al. 1993; Kaslow et al. 1987; Phair et al. 1992; Vlahov et al. 1991), which include thousands of study participants followed for up to 20 years.

The success in discovering these genes has important implications not only in describing the devastating progress of AIDS but also in developing new opportunities for therapy. Several centers are developing drugs and therapeutic protocols that emulate the genetic protection seen with these restriction genes (Cairns and D'Souza 1998, Cohen 1997; O'Brien 1998). Also, these discoveries suggest the feasibility for unraveling other complex polygenic traits as well as those where an environmental component is required for gene action assessment. Multiple genes that act at different stages of AIDS progression are clearly at work here, and we have a real notion of how they work. In addition, AIDS clearly requires an environmental component, HIV-1 exposure, effectively precluding traditional pedigree analysis in gene detection. However, the candidate gene approach spurred by functional insight from parallel disciplines plus the population genetics association approach may be a useful application for detecting loci that are polygenic, multifactorial or both.

References

Bartlett JG, Moore RD (1998) Improving HIV therapy. Sci Amer 279(1):84–87

Berger EA, Doms RW, Fenyo EM, Korber BTM, Littman DR, Moore JP, Sattentau QJ, Schuitemaker H, Sodroski J, Weiss RA (1998) A new classification for HIV-1. Nature 39:240.

Bozzette SA, Berry SH, Duan N, Frankel MR, Leibowitz AA, Lefkowitz D, Emmons CA, Senterfitt JW, Berk ML, Morton SC, Shapiro MF (1998) The care of HIV-infected adults in the United States. N Eng J Med 339:1897–1904.

Bream JH, Young HA, Rice N, Martin MP, Carrington M, O'Brien SJ (1999) CCR5 promoter alleles distinguished by specific DNA binding factors. Science 284:223a

Buchbinder SP, Katz MH, Hessol NA, O'Malley PM, Holmberg SD (1994) Long-term HIV-1 infection without immunologic progression. AIDS 8:1123–1128

Cairns JS, D'Souza MP (1998) Chemokines and HIV-1 second receptors: the therapeutic connection. Nat Med 4:563–569

Carrington M, Nelson G, Martin MP, Kissner T, Vlahov D, Goedert JJ, Kaslow R, Buchbinder S, Hoots K, O'Brien SJ (1999). *HLA* and HIV-1: Heterozygote advantage and *B*35-Cw*04* disadvantage. Science 283:1748–1752

Chan DC, Kim PS (1998) HIV entry and its inhibition. Cell 93:681–684

Chee M, Yang R, Hubbell E, Berno A, Huang XC, Stern D, Winkler J, Lockhart DJ, Morris MS, Fodor SP (1996) Accessing genetic information with high-density DNA arrays. Science 274:610–614

Cohen J (1997) Exploiting the HIV-chemokine nexus. Science 275:1261–1264

Collins FS, Guyer MS, Chakravarti A (1997) Variations on a theme: cataloging human DNA sequence variation. Science 278:1580–1581

Collins FS, Patrinos A, Jordan E, Chakravarti A, Gesteland R, Waiters L, and the members of the DOE and NIH planning groups (1998) New goals for the U.S. Human Genome Project: 1998–2003. Science 282:682–689

Collins KL, Chen BK, Kalams SA, Walker BD, Baltimore D (1998) HIV-1 Nef protein protects infected primary cells against killing by cytotoxic T lymphocytes. Nature 391:397–401

Connor RI, Ho DD (1994) Human immunodeficiency virus type 1 variants with increased replicative capacity develop during the asymptomatic stage before disease progression. J Virol 68:4400–4408

D'Souza MP, Harden VA (1996) Chemokines and HIV-1 second receptors. Nat Med 2:1293–1300

Dean M, Carrington M, Winkler C, Huttley GA, Smith MW, Allikmets R, Goedert JJ, Buchbinder SP, Vittinghoff E, Gomperts E, Donfield S, Vlahov D, Kaslow R, Saah A, Rinaldo C, Detels R, HGDS, MACS, MHCS, SF City Cohort, ALIVE Study, O'Brien SJ (1996) Genetic restriction of HIV-1 infection and progression to AIDS by a deletion allele of the *CKR5* structural gene. Science 273:1856–1862

Dean M, Jacobson LP, Mc Farlane G, Margolick JB, Jenkins FJ, Howard OM, Dong HF, Goedert JJ, Buchbinder S, Gomperts E, Vlahov D, Oppenheim JJ, O'Brien SJ, Carrington M (1999) Reduced risk of AIDS lymphoma in individuals heterozygous for the CCR5Δ32 mutation. Cancer Res 59:3561–3564.

Detels R, Liu Z, Hennessey K, Kan J, Visscher BR, Taylor JMG, Hoover DR, Rinaldo CR, Jr., Phair JP, Saah AJ, Giorgi JV (1996) For the Multicenter AIDS Cohort Study: resistance to HIV-1 infection. J AIDS 7:1263–1269

Goedert JJ, Biggar RJ, Winn DM, Mann DL, Byar DP, Strong DM, DiGioria RA, Grossman RJ, Sanchez WC, Kase RG, Greene MH, Hoover RN, Blattner WA (1985) Decreased helper T lymphocytes in homosexual man. I. Sexual contact in high-incidence areas for the acquired immunodeficiency syndrome. Am J Epidemiol 121:629–636

Goedert JJ, Kessler CM, Aledort LM, Biggar RJ, Andes WA, White GS, Drummond JE, Vaidya K, Mann DL, Eyster ME, Ragni MV, Lederman M, Cohen AR, Bray GL, Rosenberg PS, Friedman RM, Hilgartner MW, Blattner WA, Kroner B, Gail MH (1989) A prospective study of human immunodeficiency virus type 1 infection and the development of AIDS in subjects with hemophilia. N Engl J Med 321:1141–1148

Hilgartner MW, Donfield SM, Willoughby A, Contant CF, Jr, Evatt BL, Gomperts ED, Hoots WK, Jason J, Loveland KA, McKinlay SM, Stehbens JA (1993) Hemophilia Growth and Development Study. Design, methods, and entry data. Am J Pediatr Hematol Oncol 15:208–218

Ho DD, Neumann AU, Perelson AS, Chen W, Leonard JM, Markowitz M (1995) Rapid turnover of plasma virions and CD4 lymphocytes in HIV-1 infection. Nature 373:123–126

Huang Y, Paxton WA, Wolinsky SM, Neumann AU, Zhang L, He T, Kang S, Ceradini D, Jin Z, Yazdanbakhsh K, Kunstman K, Erickson D, Dragon E, Landau ER, Phair J, Ho DD, Koup RA (1996) The role of a mutant CR5 allele in HIV-1 transmission and disease progression. Nat Med 2:1240–1243

Kaslow RA, Ostrow DG, Detels R, Phair JP, Polk BF, Rinaldo CF, Jr (1987) The Multicenter AIDS Cohort Study: rationale, organization, and selected characteristics of the participants. Am J Epidemiol 126:310–318

Kostrikis LG, Huang Y, Moore JP, Wolinsky SM, Zhang L, Guo Y, Deutsch L, Phair J, Neumann AU, Ho DD (1998) A chemokine receptor CCR2 allele delays HIV-1 disease progression and is associated with a CCR5 promoter mutation. Nat Med 3:350–353

Lederman MM, Jackson JB, Kroner BL, White GC, III, Eyster ME, Aledort LM, Hilgartner MW, Kessler CM, Cohen AR, Kiger DP, Goedert JJ (1995) Human immunodeficiency virus (HIV) type 1 infection status and in vitro susceptibility to HIV infection among high-risk HIV-1-seronegative hemophiliacs. J Infect Dis 172:228–231

Lee B, Doranz BJ, Rana S, Yi Y, Mellado M, Frade JMR, Martinez-A C, O'Brien SJ, Dean M, Collman RG, Doms RW (1998) Influence of the CCR2–V64I polymorphism on human immunodeficiency virus type 1 coreceptor activity and on chemokine receptor function of CCR2b, CCR3, CCR5, and CXCR4. J Virol 72:7450–7458

Lipshutz RJ, Fodor SP, Gineras IR, Lockhart DJ (1999) High density synthetic oligonucleotide arrays. Nat Genet 21:20–24

Littman DR (1998) Chemokine receptors: keys to AIDS pathogenesis? Cell 93:677–680

Liu R, Paxton WA, Choe S, Ceradini D, Martin SR, Horuk R, MacDonald ME, Stuhlmann H, Koup RA, Landau NR (1996) Homozygous defect in HIV-1 coreceptor accounts for resistance of some multiply-exposed individuals to HIV-1 infection. Cell 86:367–377

Mann JM, Tarantola DJM (1998) HIV the global picture. Sci Amer 279(1):82–83 (United Nations Report on world demography. Released October 28, 1998)

Martin MP, Dean M, Smith MW, Gerrard B, Michael NL, Lee B, Doms RW, Margolick J, Buchbinder S, Goedert JJ, O'Brien TR, Hilgartner MW, Vlahov D, O'Brien SJ, Carrington M (1998) Genetic acceleration of AIDS progression by a promoter variant of CCR5. Science 282:1907–1911

McDermott DH, Zimmerman PA, Guignard F, Kleeberger CA, Leitman SF, Murphy PM (1998) CCR5 promoter polymorphism and HIV-1 disease progression. Multicenter AIDS Cohort Study (MACS). Lancet 352:866–870

McNicholl JM, Smith DK, Qari SH, Hodge T (1997) Host genes and HIV: the role of the chemokine receptor gene CCR5 and its allele (Δ32 CCR5). Emerg Infect Dis 3:261–271

Michael NL, Chang G, Louie LG, Mascola JR, Dondero D, Birx DL, Sheppard HW (1997) The role of viral phenotype and CCR5 gene defects in HIV-1 transmission and disease progression. Nat Med 3:338–340

Mummidi S, Ahuja SS, Gonzalez E, Anderson SA, Santiago EN, Stephan KT, Craig FE, O'Connell P, Tryon V, Clark RA, Dolan MJ, Ahuja SJ (1998) Genealogy of the CCR5 locus and chemokine system gene variants associated with altered rates of HIV-1 disease progression. Nat Med 4:786–793

O'Brien SJ (1998) A new approach to therapy. HIV Newsline 4:3–6

O'Brien SJ, Dean M (1997) In search of AIDS resistance genes. Sci Amer 277:44–51

Palella FJ, Jr., Delaney KM, Moorman AC, Loveless MO, Fuhrer J, Satten GA, Aschman DJ, Holmberg SD (1998) Declining morbidity and mortality among patients with advanced human immunodeficiency virus infection. N Engl J Med 338:853–860

Phair J, Jacobson L, Detels R, Rinaldo C, Saah A, Schrager L, Munoz A (1992) Acquired immune deficiency syndrome occurring within 5 years of infection with human immunodeficiency virus type-1: The Multicenter AIDS Cohort Study. J AIDS 5:490–496

Premack BA, Schall TJ (1996) Chemokine receptors: gateways to inflammation and infection. Nat Med 2:1174–1178

Richman DD, Bozzette SA (1994) The impact of the syncytium-inducing phenotype of human immunodeficiency virus on disease progression. J Infect Dis 169:968–974

Rosenberg ES, Billingsley JM, Caliendo AM, Boswell SL, Sax PE, Kalams SA, Walker BD (1997) Vigorous HIV-1-specific CD4+ T cell responses associated with control of viremia. Science 278:1447–1450

Samson M, Iibert F, Doranz BJ, Rucker J, Liesnard C, Farber CM, Saragosti S, Lapoumerouli C, Cognaux J, Forceille C, Muyldermans G, Verhofstede C, Burtonboy G, Georges M, Imai T, Rana S, Yi Y, Smyth RJ, Collman RG, Doms RW, Vassart G, Parmentier M (1996) Resistance to HIV-1 infection in

Caucasian individuals bearing mutant alleles of the CCR5 chemokine receptor gene. Nature 382:722–725

Schuitemaker H, Koot M, Kootstra NA, Dercksen MW, de Goede RE, van Steenwijk RP, Lange JM, Schattenkerk JK, Miedema F, Tersmette M (1992) Biological phenotype of human immunodeficiency virus type 1 clones at different stages of infection: progression of disease is associated with a shift from monocytotropic to T-cell-tropic virus population. J Virol 66:1354–1360

Smith MW, Dean M, Carrington M, Winkler C, Huttley GA, Lomb DA, Goedert JJ, O'Brien TR, Jacobson LP, Kaslow R, Buchbinder S, Vittinghoff E, Vlahov D, Hoots K, Hilgartner MW, Hemophilia Growth and Development Study, Multicenter AIDS Cohort Study, Multicenter Hemophilia Cohort Study, San Francisco City Cohort, ALIVE Study, O'Brien SJ (1997) Contrasting genetic influence of CCR2 and CCR5 receptor gene variants on HIV-1 infection and disease progression. Science 277:959–965

Vlahov D, Anthony JC, Munoz A, Margolick J, Nelson KE, Celentano DD, Solomon L, Polk BF (1991) The ALIVE Study, a longitudinal study of HIV-I infection in intravenous drug users: description of methods and characteristics of participants. NIDA Res Monogr 109:75–100

Wang DG, Fan JB, Siao CJ, Berno A, Young P, Sapolsky R, Ghandour G, Perkins N, Winchester E, Spencer J, Kruglyak L, Stein L, Hsie L, Topaloglou T, Hubbell E, Robinson E, Mittmann M, Morris MS, Shen N, Kilburn D, Rioux J, Nusbaum C, Rozen S, Hudson TJ, Lipshutz R, Chee M, Lander ES (1998) Large-scale identification, mapping, and genotyping of single-nucleotide polymorphisms in the human genome. Science 280:1077–1082

Warvick JA (1998) AIDS's shadow cools global population forecasts. Washington Post, 28 Oct. pA2

Wei X, Ghosh SK, Taylor ME, Johnson VA, Emini EA, Deutsch P, Lifson JD, Bonhoeffer S, Nowak MA, Hahn BH, Saag MS, Shaw GM (1995) Viral dynamics in human immunodeficiency virus type 1 infection. Nature 373:117–122

Winkler C, Modi W, Smith MW, Nelson GW, Wu X, Carrington M, Dean M, Honjo T, Tashiro K, Yabe D, Buchbinder S, Vittinghoff E, Goedert JJ, O'Brien TR, Jacobson LP, Detels R, Donfield S, Willoughby A, Gomperts E, Vlahov D, Phair J, ALIVE Study, Hemophilia Growth and Development Study (HGDS), Multicenter AIDS Cohort Study (MACS), Multicenter Hemophilia Cohort Study (MHCS), San Francisco City Cohort (SFCC), O'Brien SJ (1998) Genetic restriction of AIDS pathogenesis by an SDF-1 chemokine gene variant. Science 279:389–393

Zimmerman PA, Bukckler-White A, Alkhatib G, Spalding T, Kubofcik J, Combadiere C, Weissman D, Cohen O, Rubbert A, Lam G, Vaccarezza M, Kennedy PE, Kumanaswami V, Giorgi JV, Detels R, Hunter J, Chopek M, Berger EA, Fauci AS, Nutman TB, Murphy PM (1997) Inherited resistance to HIV-1 conferred by an inactivating mutation in CC chemokine receptor 5: Studies in populations with contrasting clinical phenotypes, defined racial background, and quantified risk. Mol Med 3:23–26

Gene Protecting against Age-Related Macular Degeneration

P. Amouyel

Abstract

Age-related macular degeneration (ARMD) is the most common cause of blindness in the elderly in Europe and in the United States ARMD is a multifactorial disease process corresponding to a thinning or scarring of the center of the retina, the macula. Some risk factors that have been associated with ARMD include less skin pigmentation, cigarette smoking, systemic hypertension, low intake of antioxidants and family history The main clinical retinal markers of ARMD are drusen, which are composed of protein and lipid deposits within the Bruch's membrane.

The lipid composition of drusen and the possible influence of genetic susceptibility risk factors prompted us to consider the gene coding for a protein involved in lipid transport, apolipoprotein E (APOE), as an interesting candidate risk factor for ARMD occurrence.

In a case control study, we compared patients with exudative ARMD to controls of similar age and gender. We observed a lower frequency of the *APOE* ε4 allele in patients with exudative ARMD. For non *APOE* ε4 allele bearers, the relative risk of developing exudative ARMD was almost five-fold higher than for *APOE* ε4 allele bearers.

This intriguing protective effect of the *APOE* ε4 allele, which is commonly known to be deleterious for the vascular system and cognitive performance, opens up several questions related to disease origin, gene evolution and genetic testing.

Introduction

Blindness is a major cause of disability in the world. However, the etiology of visual impairment differs in different countries. In the Western part of the world, blindness is mainly related to aging and involves alterations of the center of retina, the macula, which is the area of fine vision. This aging disease, called age-related macular degeneration (ARMD), is a multifactorial disease process wherein the macula is damaged by either thinning or scarring (Hyman 1987). Thinning corresponds to the dry or atrophic type of ARMD and scarring corresponds to the wet or exudative type. Several risk factors have been associated

V. Boulyjenkov, K. Berg, Y. Christen (Eds.)
Genes and Resistance to Diseases
© Springer-Verlag Berlin Heidelberg 2000

with ARMD, including family history of ARMD or blindness. This latter risk factor suggests the existence of a genetic component in the occurrence of ARMD.

On the basis of pathophysiology of the disease and a candidate gene approach, we focused our research on genes coding for proteins involved in lipid trafficking. In a case control study (Souied et al. 1998), we explored the impact of a key protein of lipid metabolism, the apolipoprotein E (APOE; Davignon et al. 1988). This protein exists as three common isoforms (APOE2, APOE3, APOE4) coded by three alleles (ε2, ε3, ε4, respectively) with different biophysical and biochemical properties. Moreover, these alleles have been associated with deleterious and protective effects on cardiovascular and neurodegenerative diseases (Weisgraber et al 1996) We observed a protective effect of the *APOE* ε4 allele on exudative ARMD, whereas this allele is commonly associated with an increase of myocardial infarction and Alzheimer's disease incidence. To our knowledge, this is the first report of a protective effect of the *APOE* ε4 allele.

Epidemiology of ARMD

ARMD is the most common cause of blindness in the elderly in Europe and in the United States (Hyman 1987). In Western countries, more than five million individuals aged 85 years and over will be affected by ARMD in 2020. Thus, early detection and prevention of ARMD constitute major public health problems associated with the dramatic increase in life expectancy.

The diagnosis of ARMD is made upon eye examination of patients (Bird et al. 1995). This examination, combined with fluorescein angiography, detects several lesions. Irregular pigmentation of the retina and choroid and accumulations of debris can be seen on retinal examination. There are various configurations of such debris, called drusen (from a German word meaning stony nodule). Drusen are yellow deposits in the deep part of the retina. They result from the metabolism of the retinal cells. Drusen can be subdivided into two major types: hard drusen, considered to be a common manifestation of aging, and soft drusen, which are associated with a higher risk of exudative ARMD. Other lesions such as atrophy of the retina, retinal pigment epithelial detachments and choroidal neovascularisation, can be observed.

Loss of retinal cells is a common feature of aging. Derived from the same embryological tissues as the brain, the retina does not make new cells when damaged. A normal loss of retina cells exists, due mainly to apoptosis. This normal loss allows for a healthier environment for the remaining cells, maintaining good visual function as age increases. However, some individuals will be affected by this cell loss, resulting in thinning or scarring and visual loss, usually slow and insidious.

Many factors may contribute to the normal aging process with deleterious effects in some individuals (Hyman 1987). For instance, subjects with less skin pigmentation have a higher risk of ARMD. The dark pigment of the body may absorb part of the blue light whose short wavelength reaches the macula more easily. This will produce free oxygen radicals and induce long-term cell damage. Several ocular risk factors have been reported: light exposure, hyperopia and light iris color. Systemic risk factors have also been described white race, female gender, low intake of antioxidants, cigarette smoking, systemic hypertension and family history. This last risk factor supports the hypothesis of an individual susceptibility to ARMD.

The Genetic Component of ARMD

Family history as a risk factor of ARMD is an argument for the existence of a genetic component in the occurrence of ARMD. Several observations reinforce this hypothesis, including a familial aggregation study (Seddon et al. 1997) and hereditary forms of macular degeneration (Allikmets 1997).

A recent Dutch study estimated the attributable risk related to genetic factors in ARMD (Klaver et al. 1998b). The first-degree relatives of 87 patients with late ARMD recruited within the framework of a large population-based study were compared with first-degree relatives of 135 controls without ARMD. The risk ratio to develop ARMD was 4.2 for first-degree relatives of patients. This familial aggregation study suggested that the population-attributable risk to genetic factors was 23%.

Other evidence of a genetic component in the occurrence of ARMD comes from a hereditary recessive form of macular dystrophy, Stargardt disease (estimated incidence = 1 in 10 000). This disease is characterized by juvenile-to-young adult onset, central visual impairment, progressive bilateral atrophy of the macular retinal pigment epithelium and neuroepithelium, and the frequent appearance of yellow flecks distributed around the macula and the midretinal periphery. In 1997, a photoreceptor specific gene, *ABCR* (also known as *STGD1*), located on human chromosome 1 was found to be mutated in Stargardt disease (Allikmets 1997). The *ABCR* protein is a member of the adenosine triphosphate-binding cassette (ABC) transporter superfamily, which includes active transporters of lipids, hydrophobic drugs and peptides. The phenotypic similarities of ARMD and Stargardt disease prompted the group who discovered the implication of *ABCR* in Stargardt disease to search for mutations of *ABCR* in ARMD (Allikmets et al. 1997). They screened the *ABCR* coding sequence of 167 unrelated ARMD patients and found mutations in 26 individuals (16%). Alterations of *ABCR* sequence were, with one exception, detected in the dry ARMD type. Among the 26 individuals with alterations, 50% had a first- or second-degree relative with ARMD compared to the 24% of the rest of the sample. All of these data favored a genetic susceptibility to ARMD.

Hunting for Genetic Susceptibility Risk Factors of ARMD: the Candidate Gene Approach

Two main strategies coexist for efficient gene discovery: systematic genome screening and candidate gene approach. Genome scanning requires family studies, with whole or nuclear pedigrees or with multiple pairs of affected sisters and brothers. This strategy often permits the discovery of new genes whose products may or may not be known. This strategy is useful when the existence of a genetic component, preferentially simple, is established, and when pathophysiological clues are lacking. Conversely, when transmission is complex, as for multifactorial chronic diseases, candidate gene strategies allow more rapid analyses in case-control, prospective or population studies. However, strong physiological hypotheses, internal consistencies, and independent confirmations are required.

A major component of ARMD is the drusen (Pauleikhoff et al. 1990). Drusen are seen in 50 % of the white population over 50 years of age and represent an accumulation of debris from the metabolic process of vision. These debris accumulate beneath retinal pigment epithelium through Bruch's membrane and basal lamina. Two clinical types of drusen variously associated with ARMD are described : hard drusen are a common manifestation of aging, whereas soft drusen are associated with a higher risk of exudative ARMD. Drusen, particularly soft ones, are characterized by protein and lipid deposits within Bruch's membrane. They contain cholesteryl esters, unsaturated fatty acids and sometimes phospholipids. This observation suggests that lipid trafficking may be an interesting metabolic pathway to identify a target gene.

Apolipoproteins are lipid transporters. Among these, APOE plays a central role (Davignon et al. 1988). This apolipoprotein exists as three frequent isoforms in human populations : APOE2, APOE3 and APOE4. These isoforms are associated with a heterogeneity of lipid and lipoprotein plasma levels. APOE3, the most frequent isoform (70 to 80 %), is characterized by a cystein residue at position 112 and an arginin residue at position 158 of the coding sequence, and is considered the reference isoform. APOE4, characterized by arginin residues at position 112 and 158, is associated with increased plasma concentrations of low-density lipoprotein (LDL) cholesterol levels. Conversely, APOE2, characterized by cystein residues at positions 112 and 158, is associated with lower levels of LDL-cholesterol. Moreover, APOE4 is associated with an increased risk of myocardial infarction and APOE2 is linked with a decreased risk of cardiovascular disease (Luc et al. 1994). These three isoforms are coded by three different alleles: ε2, ε3 and ε4. APOE is expressed in liver, macrophages, kidney, lung and brain. Moreover, APOE is present in senile plaques of human brains affected with Alzheimer's disease (AD). The ε4 allele was found to be a major risk factor of AD and of various other neurodegenerative processes related to cognitive impairment (Weisgraber and Mahley 1996). Similarly to what is reported for cardiovascular diseases, the ε2 allele is considered to be a major protective factor for AD. The embryonic origin of retina, the involvement of APOE in brain metabolism and nerve cell reparation processes prompted us to study the potential impact of *APOE* polymorphism in ARMD.

The *APOE* ε4 Allele as a Protective Factor for ARMD

In a case-control study comparing patients with ARMD with hard and soft drusen to controls of similar age and gender, we analyzed the influence of *APOE* alleles on the occurrence of ARMD (Souied et al. 1998). A sample of 116 unrelated patients, referred to the Eye Clinic of the University Hospital of Créteil (France) for ARMD, was recruited. Drusen were defined according to their borders (sharp for hard drusen, fuzzy for soft drusen) and to their aspects on fluorescein angiography, allowing to us to separate the cases into two groups: 77 patients with soft drusen and 39 patients with hard drusen. At least twice the number of controls of similar age (75 ± 7 years) and gender (male 35 %) were selected. The visual impairment of controls was classified by the physician in charge of their recruitment (without specific retinal examination) from all medical sources of examination and by corrected visual acuity assessment on a three-level scale. This visual disability assessment allowed us to isolate, in the control groups, a group of elderly with vision within normal limits. *APOE* genotypes were characterized from genomic DNA.

The proportion of *APOE* ε4 allele carriers was 12.1 % in ARMD versus 28.6 % in controls ($p < 0.002$). This decreased frequency of *APOE* ε4 allele carriers was even higher (7.8 %) in patients with soft drusen ($p < 0.004$), whereas it was 20.5 % in patients with hard drusen. These differences were still more pronounced when the control group was restricted to "normal vision" controls (34.1 %, $p < 0.002$ for patients with soft drusen). The corresponding allele frequencies differed significantly: 0.045 in patients with soft drusen versus 0.149 for all controls ($p < 0.003$).

The protective effect of the presence of at least one *APOE* ε4 allele for ARMD was estimated with odds ratio (OR), approximating relative risk in case-control studies. For all ARMD patients versus all controls, this OR was 0.34 (95 %CI = [0.17–0.68]); for ARMD patients with soft drusen versus controls with "normal vision," the OR was 0.16 (95 % CI = [0.05–0.43]). All these data strongly suggest that the *APOE* ε4 allele is a major potential protective genetic susceptibility risk factor for ARMD (Souied et al. 1998).

Presumption of Causality

This conclusion, based on a physiopathological hypothesis, relies on a statistical association between a frequent polymorphism and a frequent chronic disease. Presumption of causality criteria are needed to reinforce the consistency of this observation.

First, an independent study developed in a Dutch population replicated our results (Klaver et al. 1998a). A genetic association study was performed among 88 cases of ARMD and 901 controls derived from a population-based study. In this study, the *APOE* ε4 allele was associated with a decreased risk of ARMD with an OR of 0.43 (95 % CI = [0.21–0.88]).

Second, a study of APOE immunoreactivity in 15 ARMD and 10 controls found that APOE was present in basal laminar deposits and in soft drusen (Klaver et al. 1998a).

Third, two biological mechanisms, based on APOE isoform properties, may explain the protective effect of APOE4. A decline in the permeability of Bruch's membrane with increasing age has been reported (Moore et al. 1995). This membrane is a key filter for the elimination of debris resulting from metabolic processes of the retina. Failure to clear this debris may result in its accumulation in retinal layers. The presence of APOE in basal laminar deposits and in soft drusen suggests that this apolipoprotein may be a major tool for this clearance, especially for the lipid components. In contrast to APOE2 and APOE3, APOE4 lacking cystein residues is not able to form dimers. The reduced size of the protein may guarantee a better clearance of debris though a Bruch's membrane whose permeability has decreased with age. Moreover, APOE4, with two arginin residues, has more positive charges than APOE2 and APOE3, which may facilitate clearance of debris.

Impacts of *APOE* Polymorphism on Health

APOE is involved in several chronic diseases in humans. This involvement was identified through the existence of a genetic polymorphism, frequent in humans. However, depending on the type of organ implicated in the disease, the effect of each allele may vary significantly.

APOE4 is a risk factor for dyslipidemia, coronary heart disease, AD and other neurodegenerative diseases, but is a protective factor for ARMD. On the other hand, APOE2 is associated with low LDL-cholesterol levels, a lower risk of AD and better cognitive functioning. These observations are not systematically verified. For instance, all patients carrying an *APOE* ε4 allele do not develop AD. Recent studies suggest that quantitative effects (Lambert et al. 1997), associated with the qualitative effects of APOE isoforms, are important in the setting of disease risk level. These quantitative variations seem to be supported by other APOE mutations located in the promoter region (Lambert et al. 1998a, b). Moreover, interactions with environmental factors such as fat consumption or light exposure may modulate the levels of risk.

Thus, the study of a frequent polymorphism of a common protein opens new areas of knowledge in the fields of classification of diseases, molecular mechanisms involved in these pathologies, evolution and presumably in prevention and treatments. For instance, several studies suggested that vascular disease and AD may be associated more frequently than expected, and that these two pathological processes may be associated in a causative way or at least share common environmental and genetic determinants. APOE may be one of these determinants (Richard and Amouyel 1998). Identical biochemical and biophysical properties of APOE4 isoform may have various consequences, depending on the tissues where this ubiquitous protein is located: reduced affinity for LDL receptors in liver, increased affinity for amyloid peptides in brain, reduced capacity of neurite out-growth in peripheral nerves, reduced antioxidant properties, increased clearance properties in retina.

From an evolutionary point of view, the ancestral gene seems to be the ε4 allele (Hanlon and Rubinsztein 1995). If we assume that environmental pressure may boost gene evolution, the APOE4 isoform may be a "thrifty genotype rendered detrimental by progress" (Neel 1962), leading to increasing life expectancy. To our knowledge, the effect of the *APOE* ε4 allele on ARMD is the first report of a protective effect of this deleterious allele in humans. Speculating on evolution, we may suggest that, in the early ages of mammalian evolution, it was more beneficial to preserve the best eyesight, to escape predators as quickly as possible, than to preserve cognition until advanced ages, which the organism was unlikely to reach.

Finally, in the same epidemiological study, we estimated the impact of at least one *APOE* ε4 allele on the risk of severe cognitive impairment and of blindness. Almost half of the cases of cognitive impairment were attributable to the presence of at least one *APOE* ε4 allele, whereas its absence explained 55 % of the cases of legal blindness. This complexity of the impact of the APOE polymorphism still limits the interest of its use as a potential predictive test, especially without any efficient treatment or preventive strategy.

Acknowledgements

This work benefited from the fruitful and active collaboration of Dr. E. Souied, Pr. G. Coscas and Pr. G. Soubrane at the Clinique Ophtalmologique Universitaire de Créteil and of Dr. F. Richard and Dr. N. Helbecque at the INSERM U508 laboratory.

References

Allikmets R (1997) A photoreceptor cell-specific ATP-binding transporter gene (ABCR) is mutated in recessive Stargardt macular dystrophy. Nat Genet 17:122

Allikmets R, Shroyer NF, Singh N, Seddon JM, Lewis RA, Bernstein PS, Peiffer A, Zabriskie NA, Li Y, Hutchinson A, Dean M, Lupski JR, Leppert M (1997) Mutation of the Stargardt disease gene (ABCR) in age-related macular degeneration. Science 277:1805–1807

Bird AC, Bressler NM, Bressler SB, Chisholm IH, Coscas G, Davis MD, de Jong PT, Klaver CC, Klein BE, Klein R (1995) An international classification and grading system for age-related maculopathy and age-related macular degeneration. The International ARM Epidemiological Study Group. Surv Ophthalmol 39:367–374

Davignon J, Gregg RE, Sing CF (1988) Apolipoprotein E polymorphism and atherosclerosis Arteriosclerosis 8:1–21

Hanlon CS, Rubinsztein DC (1995) Arginine residues at codons 112 and 158 in the apolipoprotein E gene correspond to the ancestral state in humans. Atherosclerosis 112:85–90

Hyman L (1987) Epidemiology of eye disease in the elderly. Eye 1:330–341

Klaver CC, Kliffen M, van Duijn CM, Hofman A, Cruts M, Grobbee DE, Van Broeckhoven C, de Jong PT (1998a) Genetic association of apolipoprotein E with age-related macular degeneration. Am J Human Genet 63:200–206

Klaver CC, Wolfs RC, Assink JJ, van Duijn CM, Hofman A, de Jong PT (1998b) Genetic risk of age-related maculopathy. Population-based familial aggregation study. Arch Ophthalmol 116:1646–1651

Lambert JC, Perez-Tur J, Dupire MJ, Galasko D, Mann D, Amouyel P, Hardy J, Delacourte A, Chartier-Harlin MC (1997) Distortion of allelic expression of apolipoprotein E in Alzheimer's disease. Human Mol Genet 6:2151–2154

Lambert JC, Berr C, Pasquier F, Delacourte A, Frigard B, Cottel D, Perez-Tur J, Mouroux V, Mohr M, Cecyre D, Galasko D, Lendon C, Poirier J, Hardy J, Mann D, Amouyel P, Chartier-Harlin MC (1998a) Pronounced impact of Th1/E47cs mutation compared with -491 AT mutation on neural APOE gene expression and risk of developing Alzheimer's disease. Human Mol Genet 7:1511–1516

Lambert JC, Pasquier F, Cottel D, Frigard B, Amouyel P, Chartier-Harlin MC (1998b) A new polymorphism in the APOE promoter associated with risk of developing Alzheimer's disease. Human Mol Genet 7:533–540

Luc G, Bard JM, Arveiler D, Evans A, Cambou JP, Bingham A, Amouyel P, Schaffer P, Ruidavets JB, Cambien F (1994) Impact of apolipoprotein E polymorphism on lipoproteins and risk of myocardial infarction. The ECTIM Study. Arterioscler Thromb 14:1412–1419

Moore DJ, Hussain AA, Marshall J (1995) Age-related variation in the hydraulic conductivity of Bruch's membrane. Invest Ophthalmol Vis Sci 36:1290–1297

Neel JV (1962) Diabetes mellitus: a thrifty genotype rendered detrimental by progress. Am J Hum Genet 14:353–362

Pauleikhoff D, Barondes MJ, Minassian D, Chisholm IH, Bird AC (1990) Drusen as risk factors in age-related macular disease. Am J Ophthalmol 109:38–43

Richard F, Amouyel P (1998) Genetical links between stroke and Alzheimer's disease. In: Leys D, Pasquier F, Scheltens P (eds) Current issues in neurodegenerative diseases, Vol 9, Stroke and Alzheimer's disease, ICG The Hague, pp 112–123

Seddon JM, Ajani UA, Mitchell BD (1997) Familial aggregation of age-related maculopathy. Am J Ophthalmol 123:199–206

Souied EH, Benlian P, Amouyel P, Feingold J, Lagarde JP, Munnich A, Kaplan J, Coscas G, Soubrane G (1998) The ε4 allele of the apolipoprotein E gene as a potential protective factor for exudative age-related macular degeneration. Am J Ophthalmol 125:353–359

Weisgraber KH, Mahley RW (1996) Human apolipoprotein E: the Alzheimer's disease connection. FASEB J 10:1485–1494

Genes Involved in Resistance to Carcinogenesis

C. R. Wolf

Introduction

The pathogenesis of human disease is influenced by many interacting factors and is determined by the balance between the environment in which we live and genetic susceptibility (Doll and Peto 1981). This is exemplified by the variations in disease incidence in different parts of the world. For example, migration studies show that, whereas breast cancer incidence in Japan is low, in Japanese women living in the USA it is similar to that of the rest of the US population. This demonstrates that environmental factors are the principal determinants of disease incidence (Potter 1997). However, within any population the observation that certain individuals are more susceptible than others to a particular disease indicates that genetic factors are important.

A large number of environmental factors have been implicated in disease pathogenesis. These include infectious agents as well as environmental chemicals. Toxic chemicals are produced by numerous organisms as part of their natural defence against predators or to gain selective advantage in a particular nutrient environment. Indeed, many of the chemicals produced naturally are far more toxic than the chemical agents produced as a consequence of our industrial environment. For example, many of the cytotoxic drugs that are successfully used in the treatment of cancer are natural products. As part of the evolutionary process, therefore, all organisms are constantly exposed to chemical agents that may be toxic to them; their capacity to withstand this toxic insult is fundamental to their survival (Hayes and Wolf 1997).

A large number of genes have specifically evolved to protect against chemical toxicity; many of these are conserved across phylogeny. The relative levels of expression of these genes determine our sensitivity or resistance to environmental agents and, therefore, individuality in their expression will be an important determinant of disease susceptibility.

Protection against toxic agents in multicellular organisms is complex and can occur at several levels (Hayes and Wolf 1990). The primary barrier is absorption through the lungs, skin or gastrointestinal (GI) tract. It is now known that the GI tract contains a number of enzymes involved in xenobiotic metabolism and transport (Lown et al. 1997; Watkins 1997). These proteins include drug transporters such as the multi-drug resistance protein, MDR1, and members of the cytochrome P450-dependent monooxygenase system (Watkins 1997). The next

V. Boulyjenkov, K. Berg, Y. Christen (Eds.)
Genes and Resistance to Diseases
© Springer-Verlag Berlin Heidelberg 2000

line of defence is hepatic metabolism. The uptake and elimination of chemicals into and from the hepatocytes are again determined by drug transporter proteins and, most importantly, by drug metabolising enzymes. Hepatic drug metabolising enzymes are the major sites of xenobiotic metabolism in mammals (Gibson and Skett 1994). The relative levels of these enzymes will establish the circulating level of drug or toxin within the body.

The sensitisation of a target cell to a toxin or mutagen can also be determined by the activity of drug transporters and the capacity to inactivate the xenobiotic through metabolism. However, proteins which repair damage and prevent cell death by cell cycle arrest can also play a major role. Some exciting recent advances have identified cell signalling and sensor pathways that react to toxic stress. These include the MAP kinase and SAP kinase signalling pathways that activate transcription factors such as fos and jun and also regulate cellular response to DNA damage through the actions of genes such as p53 (Downes et al. 1998). These pathways determine, for example, whether a cell survives or dies by apoptosis. Therefore, a large number of genes within the mammalian genome have the capacity to influence the sensitivity of a cell to a particular type of toxic challenge. The properties of some of these genes are described below.

Individuality in Response to Toxins and Carcinogens

It can be hypothesised that genetic variability in the level or function of the proteins involved in cytoprotection will result in altered sensitivity to environmental agents and therefore be important in susceptibility to disease. In this article, the discussion will focus on the genes involved in the metabolism and disposition of xenobiotic chemicals. Genetic studies on polymorphisms in other metabolic pathways, apart from genes which involve a strong familial risk, remain to be carried out in detail.

The enzymes involved in foreign compound metabolism have been trivially termed "drug metabolising enzymes". This term belies their evolutionary importance and the central role of these enzyme systems in the life and death of cells (Wolf 1986; Gonzalez and Nebert 1990). However, this definition does identify another important function of these proteins, i.e., their role in the metabolism and disposition of therapeutic drugs. Therefore, individuality in their expression is also an important determinant of the outcome of drug therapy.

Nearly all lipophilic small chemical molecules which enter the body have to be metabolised to more polar products before they can be eliminated. Indeed, certain types of industrially produced chemicals, such as the polychlorinated biphenyls, are inert to metabolism and can stay in the body for many years in fat tissue. This fact is demonstrated by the studies which show that such chemicals can be eliminated in high concentration in milk fat (Borlakoglu et al. 1993). In the pioneering work of RT Williams in the late 1950s, hepatic drug metabolising enzymes were classified into two groups, termed Phase 1 and Phase 2 (Williams 1959). Phase 1 enzymes catalyse "functionalisation reactions" involving the insertion of an atom of molecular oxygen into a C–H bond to generate a hydroxylated product. Such reactions are almost exclusively catalysed by the cytochrome P450-dependent monooxygenase system. The metabolites of P450-mediated metabolism are further conjugated to glutathione, glucuronic acid or sulphate by enzymes such as the glutathione S-transferases, UDP-glucoronyl transferases or sulphotransferases to further increase polarity and facilitate elimination (Gibson and Skett 1994).

Cytochrome P450-catalysed monooxygenase reactions (Nelson et al. 1996) are highly energetic and can result in the conversion of a foreign chemical to a product which is electrophilic and more toxic or mutagenic than the parent compound. This reaction pathway is epitomised by the activation of the polycyclic aromatic hydrocarbon lung carcinogens present in cigarette smoke, the activation of the analgesic paracetamol to hepatic products and the metabolism of the mycotoxin Aflatoxin B1 to a highly mutagenic epoxide intermediate. Metabolism of Aflatoxin B1 to products which form DNA adducts is thought to be a major contributing factor in the incidence of liver cancer in Africa and Asia (Potter 1997; Fig. 1). The mutagenic or toxic effects of these highly reactive intermediates can be countered by the action of Phase 2 enzymes such as epoxide hydrolase or the glutathione S-transferases. Probably the primary mechanism of detoxification of such reactive intermediates is through conjugation with glutathione catalysed by the glutathione S-transferases. Therefore, the relative level of specific cytochrome P450s or glutathione S-transferases within cells can determine both the rate of detoxification of a particular harmful chemical agent and the level of a cytotoxic or mutagenic metabolite. The generation of such metabolites is an essential prerequisite for chemical carcinogenesis induced by mutagens in man.

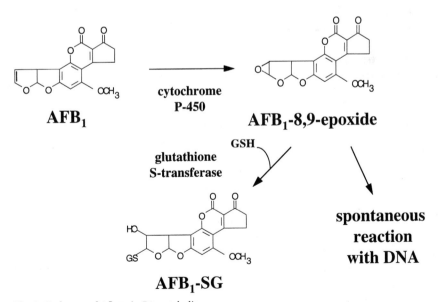

Fig. 1. Pathways of Aflatoxin B1 metabolism.

The Cytochrome P450-dependent Monooxygenases

It has been estimated that the cytochrome P450-dependent monooxygenase family evolved over one thousand million years ago (Gonzales and Nebert 1990). It can be hypothesised that the original function of these enzymes was to use molecular oxygen to convert hydrocarbons to hydroxylated products which could then be used as an energy source (Wolf 1986). Indeed, many organisms which grow on hydrocarbons are able to do so because of the presence of cytochrome P450 enzymes. Subsequently, this enzyme system has evolved into a multigene family of proteins which carry out a wide range of critical functions within mammals. (Fig. 2). P450 proteins are involved in the biosynthesis of all steroid hormones, vitamin D and retinoic acid, as well as in the production of bile acids from cholesterol. In addition to these constitutive functions, a large number of P450s have specifically evolved for the metabolism and disposition of foreign compounds.

To date, approximately 20 human cytochrome P450s have been identified which have the capacity to metabolise drugs and foreign compounds (Fig. 3). These enzymes are part of a supergene family, subdivided into families and subfamilies based on sequence homology and chromosomal location. Each individ-

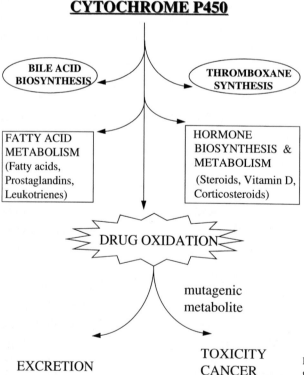

Fig. 2. Reactions catalysed by the cytochrome P450 system.

Gene		Chromosomal localisation	Tissue distribution
CYP1	CYP1A1	15q22-q24	extra-hepatic
	CYP1A2	15q22-qter	liver
	CYP1B1	2p22-p21	extra-hepatic
CYP2	CYP2A6	19q13,2-13.1	liver, testis
	CYP2A7	"	-
	CYP2A13	"	-
	CYP2B6	19q12-13.2	liver, lung
	CYP2C8	10q24.1-24.3	liver
	CYP2C9	"	"
	CYP2C18	"	"
	CYP2C19	"	"
	CYP2D6	22q13.1	liver, kidney
	CYP2E1	10q24.3-qter	liver, pancreas
CYP3	CYP3A4	7q22-qter	liver, GI tract
	CYP3A5	"	"
	CYP3A7	"	foetal liver

Fig. 3. Cytochrome P450 gene families.

ual cytochrome P450 isozyme has a distinct but, in some cases, overlapping substrate specificity with other family members. Most P450s are expressed predominantly in the liver, but many are also expressed in several extrahepatic tissues. Certain P450 enzymes are expressed almost exclusively in extrahepatic tissues.

Analysis of cytochrome P450 gene expression across a panel of human liver samples has demonstrated that the levels of these proteins are subject to very large inter-individual variation. This variation can be more than 60-fold for certain P450 forms (Forrester et al. 1992; Shimada et al. 1994). This variability in expression can be either environmentally or genetically controlled (see below). Quantification of the mean cytochrome P450 content for different isozymes across a typical liver panel and the relative variability in expression is shown in Table 1. The major human hepatic cytochrome P450 is CYP3A4. It is interesting to note that one of the minor forms, CYP2D6, which accounts for only 1.5 % of the total hepatic P450 content, plays a major role in the metabolism and disposition of drugs in man. A list of examples of some substrates for CYP3A4 and CYP2D6 is shown in Table 2. It has been estimated that cytochrome P450 CYP2D6 is responsible for the metabolism of up to 25 % of therapeutic drugs. The reason why CYP2D6 plays such an important role in the disposition of therapeutic drugs appears to be because a large proportion of drugs are targeted to the central nervous system. Most of these drugs require a basic nitrogen atom for pharmacological activity, which is also a key structural feature of CYP2D6 substrates (Wolff et al. 1985; Islam et al. 1991). CYP2D6 is found in the substantia nigra of the brain and it would appear that it has evolved to inactivate environmental neurotoxins (Coleman et al. 1996; Gilham et al. 1997).

Table 1. Inter-individual variation in hepatic cytochrome P450 content

P450	Specific content[1] (pmol/mg protein)	Fold variation[2]	% drugs metabolised
CYP1A2	42 ± 23	10	5
CYP2A6	14 ± 13	13	<1
CYP2B6	1 ± 2	nd[3]	<1
CYP2C	60 ± 27	CYP2C8 60 CYP2C9 3	19
CYP2D6	5 ± 4	16	24
CYP2E1	22 ± 12	4	1
CYP3A	96 ± 51	60	51

[1] Adapted from Forrester et al. 1992
[2] Adapted from Shimada et al. 1994
[3] nd, not determined

Table 2. Some cytochrome P450 substrates

CYP3A4 substrates

Acetaminophen	Diazepam	Loratadine	Sulfentanil
Aflatoxin B1	Digitoxin	Losartan	Steroid → 6-beta
Aldrin	Dihydroergotamine	Lovastatin	Tacrolimus (FK-506)
Alfentanil	Diltiazem	MDL 73005	Tamoxifen
Amiodarone	Doxorubicin	Midazolam	Taxol
6-Aminochrysene	Ebastine	Mitoxantrone	Teniposide
Astemizole	Ergot CQA 206–291	Navelbine	Terfenadine
Bayer R4407/5417	Erythromycin	Nifedipine	Testosterone
Benzopyrene	Estradiol	1-Nitropyrene	Tetrahydrocannabinol
Benzphetamine	Ethynylestradiol	Omeprazole	Theophylline
Budenoside	Etoposide	Ondansetron	Toremifene
Carbamazepine	Felodipine	Pot. Canrenoate	Troleandomycin
Cocaine	Flutamine	Paracetamol	Triazolam
Codeine	Folinic Acid	Proguanil	S-Warfarin
Cortisol	Gestodene	Propafenone	Verapamil
CQA 206–291	Hydroxyarginine	Quinidine	Vinblastine
Cyclophosphamide	Imipramine	Rapamycin	Vincristine
Cyclosporin A & G	Lansoprazole	Retinoic Acid	Vindesine
Dapsone	Lidocaine	Salmetorol	Zatosetron
Dextromethorphan	Lomustine	Sertindole	Zonisamide
DHEA	L-696229	Sulphamethoxazole	

Table 2. (Continue)

CYP2D6 substrates

Alprenolol	Encainide	Methoxypsoralen	N-Propylajmaline
Amiflavine	Ethinylestradiol	Methysergide Hcl	Propafenone
Amiodorone	Ethylmorphine	Metoclopramide	Propranolol
Amitryptiline	Fenoterol	Metoprolol	Pyrimethamine
Apigenin	Flecainide	Minaprine	Quercitin
Budesonide	Fluvoxamine maleate	Moclobemide	Rifampicin
Bufuralol	Fluoxetine	MPTP	Ritonavir
Bupranolol	Formoterol	Mexiletine	Roxithromycin
Chloral hydrate	Guanoxan	Nicergoline	Serotonin
Clomipramine	Haloperidol	Nimodipine	Sparteine
Clonidine	4-hydroxy	Nitrendipine	Sulfasalazine
Clotrimazole	amphetamine	Nortriptyline	Tacrine
Clozapine	Imipramine	Olanzapine	Tamoxifen
Codeine	Indoramine	Ondansetron	Thioridazine
Cyclobenzaprine	Ketoconazole	Oxprenolol	Timolol
Desipramine	Laudanosine	Paroxetine	Tomoxetine
Dexfenfluramine	Levomepromazine	Perhexiline	Tranylcypromine
Dextromethrophan	Loratadine	Perphenazine	Tropisetron
Dibucaine	MDMA (ecstacy)	Phenformin	Zuclopenthixol
Dihydroergotamine	Mefloquine	Phenylpropanolamine	
Dolasetron	Methoxamine Hcl	Procainamide	
Doxorubicin	Methoxyphenamine	Promethazine	

Variability in cytochrome P450 gene expression can be due to either genetic, environmental or hormonal factors. This level of variability can result in profound individual differences in sensitivity to drugs, toxins and mutagens. The cytochrome P450 system, like the immune system, is an adaptive response to environmental challenge. Exposure to a particular environmental agent often results in the induction of a specific hepatic P450 enzyme involved in its metabolism and elimination. One example of this form of regulation is the induction of cytochrome P450s in the CYP1A gene family through the action of the Ah receptor. The Ah receptor is a transcription factor which is retained in the cytoplasm complexed with heat shock protein 90 (Hsp 90). This complex is dissociated when small molecule inducing agents such as polycyclic aromatic hydrocarbons bind to the receptor and allow its translocation into the nucleus. Within the nucleus, the Ah receptor dimerises with the Ah receptor nuclear translocator protein (ARNT) and activates the promoters of responsive genes such as cytochrome P450 CYP1A1 (Nebert et al. 1993; Reisz-Porszasz et al. 1994).

To identify the agents which regulate gene expression by this mechanism and highlight the tissues and cell types responsive to this mode of gene regulation, we have made transgenic mice containing a 9 kilobase fragment of the rat CYP1A1 promoter linked to the *LacZ* reporter gene. Mice containing this reporter system exhibit no background *LacZ* activity. However, on administration of Ah receptor ligands a profound increase in reporter activity is observed in the liver and in many other tissues (Campbell et al. 1996; Fig. 4). This demonstrates the power of

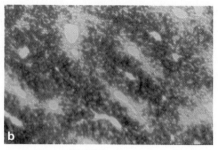

Fig. 4. Hepatic regulation of the CYP1A1 cytochrome P450 CYP1A1 promoter in transgenic mice. Transgenic mice were created containing a 9 kb fragment of the CYP1A1 promoter linked to the *LacZ* reporter. A = *LacZ* activity in the liver of untreated transgenic mice. B = *LacZ* activity in mice treated with 3-methylcholanthrene.

this adaptive response system to environmental challenge and highlights the important role it plays in our defence against toxic agents.

In addition to providing a model to understand patterns of gene regulation through the Ah receptor, the extremely tight on/off regulation of this and related promoters makes them ideal for the conditional regulation of gene expression in transgenic animals or in gene therapy. Part of the strength of this expression system is that the level of transgene expression can be "fine-tuned" by the use of inducing agents with different binding affinities and pharmacokinetics. It should therefore be possible to regulate both the intensity and duration of transgene induction using this system.

The Ah receptor mechanism provides one example of the regulation of cytochrome P450 gene expression by small molecules. Significant advances have also recently been made in the regulation of genes in other families or subfamilies of P450 genes, including the CYP2B, CYP3A and CYP4A gene families (Table 3). It is important to note that many other genes which play a chemoprotective role are also regulated by these, or similar, mechanisms. These genes include the glutathione S-transferases, glucuronyl transferases and epoxide hydrolase. Certain of these enzymes are also regulated by additional pathways through AP1 or the antioxidant responsive element (ARE). The ARE has been shown to bind the transcription factors Nrf2 and small Maf proteins (Kerppola and Curran 1994). Induction of the expression of genes by this mechanism is a major mechanism of action of chemoprotective antioxidants such as butylated hydroxyanisole and components of fruit and vegetables such as indoles (Hayes and Pulford 1995).

Individual sensitivity to chemical toxins and mutagens can therefore arise as a consequence of differences in the levels of cytoprotective genes regulated by the above mechanisms. However, there is an increasing body of evidence indicating that genetic variability in the level of expression of these proteins and the presence of allelic variants with different activities is also important. Much of the work demonstrating that genetic variability has phenotypic consequences has come from studies on the metabolism and disposition of therapeutic drugs, i.e.,

Table 3. Some receptors mediating the induction of drug metabolising enzymes

Transcription factors	Promotor element	Gene family regulated	Inducing agents
Ah receptor[1] ARNT[2]	Xenobiotic response element (XRE)	CYP1A1/CYP1B1 CYP1A2 Quinone Reductase Glucuronyl transferase UGT1A6	Polychlorinated biphenyls, Polycyclic aromatic hydrocarbons, Dioxins β-Naphthoflavone, Indole-3-carbinol etc
Not known NFI[3] CAR* cEBP	Barbie box	CYP2B GSTA Glucuronyl transferase (UGT1A1) UGT2B7	Barbiturates, TCPOBOP, DDT, Polychlorinated bipenyls Trans-stilbene oxide
PPAR α*[4] RXR*	Peroxisone proliferator response element (PPRE)	CYP4A, Glucuronyl transferase UGT1A1	Peroxisome proliferators Fatty acids
PXR*[5]	Progesterone response element (PRE)	CYP3A	TCPOBOP Rifampicin Dexamethasone Pregnenolone 16α-carbonitrile
HNF4*/Coup*[6]		CYP2C, CYP2D subfamily members	None known
NRF2/MAF[7]	Antioxidant response element (ARE)	GSTA (GSTP) Quinone Reductase Dihydrodiol dehydrogenase α GCS, UGT1A6	Indole-3-carbinole, Butylated hydroxyanisole β-Naphthoflavone, t-Butylhydroquinone Oltipraz etc
Fos/Jun[8, 9]	AP1 TPA responsive Element (TRE)	GSTP Quinone Reductase	Chemicals which induce oxidative stress? Cytokines?

* Nuclear receptors
[1] Nebert et al. 1993
[2] Reisz-Porszasz et al. 1994
[3] Sueyoshi et al. 1999
[4] Issemann and Green 1990
[5] Lehmann et al. 1998
[6] Chen et al. 1994
[7] Rushmore et al. 1990
[8] Kerppola and Curran 1994
[9] Hayes and Pulford 1995

the study of pharmacogenetics (Wolf and Smith 1999; Kalow 1992). Individuality in response to drugs and idiosyncratic adverse drug reactions have always been a problem in therapeutics. The appreciation that this individuality may be genetically based was first noted in the 1950s. Genetic polymorphism in the expression of cytochrome P450 genes was discovered in the 1960s and is perhaps now best exemplified by the polymorphism at the cytochrome P450 CYP2D6 gene locus.

This polymorphism was identified when individuals receiving the antihypertensive drug, debrisoquine, or the antiarrythmic drug, sparteine (Kalow 1992), exhibited serious side effects and abnormal drug pharmacokinetics. These effects were then ascribed to inherited abnormalities in the expression of cytochrome P450 CYP2D6. More recently, the cDNA encoding this protein has been isolated (Gonzalez et al. 1988); the molecular basis of this genetic polymorphism established and DNA based tests developed to identify individuals with impaired CYP2D6 metabolism (Gough et al. 1990). Many alleles at the cytochrome P450 CYP2D6 gene locus have now been identified (Daly et al. 1996). Intriguingly, one allelic variant involves the amplification of the entire CYP2D6 gene, with certain individuals inheriting up to 13 active copies of CYP2D6 (Johansson et al. 1993). This genotype results in an ultrarapid metaboliser phenotype, with the consequence that some drugs no longer have any therapeutic effect. Gene inactivating mutations at the CYP2D6 locus affect approximately 6 % of the Caucasian population, i.e., in Europe alone 15 million people have a compromised ability to metabolise a large number of therapeutic drugs, many of which have a narrow therapeutic index.

Another CYP2D6 allele, found predominantly in Oriental populations, results in a slower metaboliser phenotype. As a consequence, a large proportion of the Chinese population cannot tolerate the same level of drugs which are CYP2D6 substrates as Caucasians. Interethnic variation is therefore another important aspect of these enzyme systems. The amplification of the CYP2D6 gene is found in 20 % of the Ethiopian population, 7 % of the Spanish population and only 1 % of Northern Europeans (Daly et al. 1996). This interesting observation also gives us some insight into how this polymorphism became disseminated. In this case, it would appear that the gene amplification first originated in central Africa and, through migration, moved into southern Spain and then into northern Europe.

The study of pharmacogenetics is an increasingly important aspect of drug development. The ability to identify individuals at risk also raises the possibility of individualising drug treatment according to genotype as it identifies individuals who may be hypersensitive to a new drug treatment or, conversely, who may not respond (Wolf and Smith 1999). In relation to disease susceptibility, pharmacogenetic studies have proved to be very valuable as they have demonstrated that genetic polymorphisms at a particular gene locus cause a phenotypic alteration in sensitivity to chemicals *in vivo*. From this, it is reasonable to hypothesise that individuality in the expression of these genes will lead to differences in our response to environmental toxins and mutagens. The question therefore arises of how important specific genes or their allelic variants are in determining susceptibility to cancer.

To establish whether a single gene is sufficiently important to alter cancer susceptibility, we have studied the role of glutathione S-transferase pi in tumorigenesis. Glutathione S-transferase pi (GSTP) is overexpressed in cell lines made resistant to anticancer drugs, in drug-resistant preneoplastic foci in rat liver and also in a wide range of human tumours (Black and Wolf 1991; Bammler et al. 1994). These data indicate that GSTP is an important determinant of cellular defence against toxic agents. To clearly establish its role, we have deleted both GSTP genes from the mouse genome. Unlike any other species studied to date, there are two murine GSTP genes, Gstp1 and Gstp2 (Bammler et al. 1994). Both genes were inactivated by homologous recombination, resulting in the complete deletion of Gstp1 and the partial deletion of Gstp2. GSTP null mice exhibited a normal phenotype, indicating that these genes were not important in development or survival. Western and Northern blot analyses of the null mice demonstrated that the expression of both GSTP genes had been successfully deleted. Also, the metabolism of the GSTP-specific substrate, ethacrynic acid, was no longer detectable in the null animals. The metabolism of the general GST substrate, 1-chloro-2,4-dinitrobenzene, was essentially unchanged in the liver, however, indicating that other GST proteins play a significant role in the metabolism of this compound. In the lung, where GSTP is the major GST, a significant reduction in CDNB activity was observed.

To investigate whether the GSTP null animals exhibited altered sensitivity to chemical carcinogen, tumour incidence in the skin of mice was determined following exposure to the polycyclic aromatic hydrocarbon, dimethylbenz(a,h)anthracene. The results of this study are shown in Figure 5 (Henderson et al. 1998). A marked difference in sensitivity between the null and control animals was observed, in both the time taken for papilloma formation and in the numbers of papillomas formed. A three to four-fold increase in the number of papillomas in the GSTP null animals was observed.

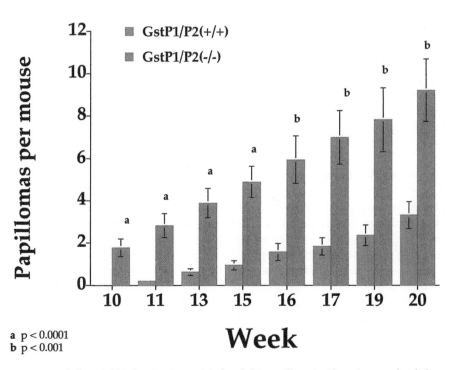

Fig. 5. Dimethylbenz(a,h)anthracene (DMBA)-induced skin papilloma incidence in control and glutathione S-transferase GSTP1 null mice. DMBA was applied to the skin of control and GSTP1 mice followed by the tumour-promoting agent, TPA.

These data demonstrated that a single gene, GSTP, can determine sensitivity to the carcinogenic effects of polycyclic aromatic hydrocarbons. These compounds are the major carcinogenic components of cigarette smoke. As GSTP is the predominant GST expressed in the lung, it is possible that variability in the expression of this gene may be an important factor in lung cancer susceptibility. Studies are currently being carried out to investigate this possibility. It is interesting to note that, in subsequent work, we have investigated the sensitivity of the GSTP null animals to chemical hepatotoxins such as paracetamol (acetaminophen; Henderson et al., unpublished). This compound is activated by the cytochrome P450 system to a chemically reactive intermediate that causes liver necrosis. This reactive intermediate can be detoxified via a glutathione-dependent mechanism through spontaneous reaction with glutathione or in an enzymatic step catalysed by the glutathione S-tranferases. GSTP has been identified as the most active GST in catalysing this reaction. Intriguingly, exposure to paracetamol resulted in a marked resistance, rather than sensitisation, to the hepatotoxic effects of this compound. Concomitant with this observation, a marked depletion of hepatic glutathione levels was observed in the drug-sensitive, wild type animals. Glutathione levels in the GSTP nulls remained unchanged. At the present time, the mechanism of this unexpected finding is unclear. However, these data demonstrate that GSTP can have a marked effect on the sensitivity of cells to toxic chemical agents activated by the P450 system. It therefore appears that this gene may play an important role in determining the sensitivity of cells to environmental agents.

In addition, it has recently been demonstrated that GSTP1 is an inhibitor of the stress-activated jun kinase pathway (Adler et al. 1999). This pathway can switch cells down chemoprotective or apoptotic pathways depending on the circumstances (Jarpe et al. 1998). It is feasible that the influence of GSTP on JNK may provide an explanation for the results observed with paracetamol.

Carcinogen Metabolising Enzymes and Cancer Susceptibility

The above data suggest that individuality in the levels of drug or carcinogen metabolising enzymes can affect cancer susceptibility. There have now been a large number of studies to establish whether genetic polymorphism in the genes encoding these proteins are susceptibility factors. By their nature, these are difficult studies because of the complexities of gene-environment interactions. The presence of a polymorphic gene will not generate susceptibility unless there is an exposure to specific causative agents metabolised by that enzyme. The same disease may be caused by a variety of environmental factors. For example, the high incidence of kidney damage and kidney disease in areas of the Balkans has been ascribed to exposure to the potent nephrotoxin ochratoxin A, produced by a mould which grows specifically in that region. A polymorphic gene which affects sensitivity to ochratoxin A would therefore not be relevant to susceptibility to kidney damage or kidney cancer in other parts of the world where this exposure does not occur. The same case can be made for many other types of cancer, such as liver cancer in Africa and Asia induced by Aflatoxin B1. Indeed, even with Europe there is sufficient variability in both diet and lifestyle to complicate the analysis of cancer susceptibility studies.

Superimposed upon these confounding factors is the observation that allelic variants are subject to different global distributions. Therefore, a polymorphism that may be important in one population may not be present in another and cannot be a susceptibility gene. This may increase the importance of one gene relative to another. The epidemiology on cancer susceptibility has been summarised in a number of review articles (Wolf 1990; Smith et al. 1995), and is the subject of

Table 4. Examples of associations between pharmacogenetic polymorphisms and disease susceptibility

Gene	Allele	Disease association	OR (95 % CI)	Population[1]
CYP1A1	$CYP1A1_{6235C}$	Lung cancer	2.03 (1.03–4.01)	USA
			2.16 (0.96–5.11)	Germany
		Bladder cancer	0.67 (0.33–1.39)	Germany
	$CYP1A1_{462Val}$	Lung cancer	2.26 (0.82–6.26)	USA
			3.01 (1.29–7.26)	Germany
		Bladder cancer	0.92 (0.61–1.41)	Germany
CYP2D6	$CYP2D6_{null}$	Skin cancer (BCC)	1.266 $(0.001)^2$	UK
		Leukaemia	12.90[3] $(0.001)^2$	UK
		Parkinson's disease	2.54 (1.61–4.28)	UK
GSTM1	GSTM1null	Lung cancer	1.4 (0.74–2.61)	Spain
			1.33 (0.89–1.99)	Norway
			1.1 (0.7–1.8)	France
			1.17 (0.98–1.4)	Caucasian (meta-analysis)
		Breast cancer	2.1 (1.22–3.64)	USA
		Bladder cancer	1.4 (1.02–1.92)	Germany
CYP1A1/ GSTM1	$CYP1A1_{6235C}$ $GSTM1_{null}$	Lung cancer	3 (1.2–7.2)	Sweden
			16 (3.76–68.02)	Japan
CYP1A1/ GSTM1	$CYP1A1_{462Val}$ $GSTM1_{null}$	Lung cancer	41 (8.68–193.61)	Japan
GSTP1	GSTP1b	Testucular cancer	3.3 (1.5–7.7)	UK
		Bladder cancer	4.2 (1.4–12.4)	UK
		Lung cancer	1.7 (1.13–2.57)	Denmark
NAT2	Slow acetylator	Bladder cancer	1.32 (0.97–1.8)	Germany
			3.19 (1.02–9.99)	USA
			1.32 (0.91–1.92)	USA
	Rapid acetylator	Colorectal cancer	2.8 (1.4–5.7)	USA
			1.29 (0.9–1.84)	USA

[1] Wolf and Smith 1999; Smith et al. 1995; Bonfetta 1999

[2] p-value

[3] x^2 value

a book to be published shortly (Bonfetta 1999). In the case of cytochrome P450 CYP2D6, there is some evidence that polymorphism in this gene may be linked to susceptibility to skin and bladder cancer and possibly to leukaemia (Table 4). However, the evidence to date is not very strong. The most interesting association of the cytochrome P450 CYP2D6 polymorphism remains that with Parkinson's disease (Smith et al. 1992). CYP2D6 is known to metabolise compounds which cause Parkinson's-like symptoms in man; CYP2D6 has also been shown to be localised in the susceptible cells in the substantia nigra of the brain (Nebert et al. 1993).

Perhaps the most encouraging results on associations between pharmacogenetic polymorphisms and disease susceptibility come from work with the glutathione S-transferases and the N-acetyl transferases. There is some evidence that polymorphism at the glutathione S-transferase GSTP1 locus may be associated with susceptibility to lung cancer, and that the prevalence of particular alleles within the cancer population is also associated with increased levels of carcinogen DNA adducts (Ryberg et al. 1997). There is also a growing body of evidence that polymorphisms in the glutathione S-transferase genes GSTP1 and GSTM1 are associated with specific cancer types, including lung cancer and leukaemia. However, the evidence to date linking specific pharmacogenetic polymorphisms with cancer susceptibility remains contradictory. It is important that further studies are carried out using well-defined cases and controls within specific populations so that initial positive or negative associations can be validated.

In conclusion, individuality in the enzyme systems involved in drug and carcinogen metabolism are extremely important determinants of human health, not only because they may determine susceptibility to disease but also because they play a central role in the metabolism and disposition of therapeutic drugs. Individuality in the expression of these genes can be both metabolically and genetically controlled. Understanding the relative role of genetic polymorphism in determining susceptibility to cancer and to a variety of other diseases remains an important theme for future studies.

Acknowledgements

I would like to thank Dr. Gillian Smith for her great help with preparing the manuscript and Dr. Sandra Campbell and Dr. Colin Henderson for Figures 4 and 5.

References

Adler V, Yin Z, Serge YF, Benezra M, Rosario L, Tew KD, Pincus MR, Sardana M, Henderson CJ, Wolf CR, Davis RJ, Ronal Z (1999) Regulation of JNK signalling by GSTp. EMBO J 18:1321–1334

Bammler TK, Smith CAD, Wolf CR (1994) Isolation and characterisation of two mouse Pi class glutathione S-transferase genes. Biochem J 298:385–390

Black SM, Wolf CR (1991) The role of glutathione-dependent enzymes in drug resistance. Pharmac Ther 51:139–154

Bonfetta P (ed) (1999) Metabolic polymorphisms and cancer. IARC Publications, France

Borlakoglu JT, Henderson CJ, Wolf CR (1993) Lactational transfer of 2,4,5', 4', 5'-hexachlorobiphenyl but not 3,4,3', 4'-tetrachlorobiphenyl, induces neonatal CYP4A1. Biochem Pharmacol 45:769–771

Campbell SJ, Carlotti F, Hall PA, Clark AJ, Wolf CR (1996) Regulation of the CYP1A1 promoter in transgenic mice: an exquisitely sensitive on-off system for cell specific gene regulation. J Cell Sci 109:2619–2625

Chen D, Lepar G, Kemper B (1994) A transcriptional regulatory element common to a large family of hepatic cytochrome P450 genes is a functional binding site of the orphan receptor HNF-4. J Biol Chem 269:5420–5427

Coleman T, Ellis SW, Martin IJ, Lennard MS, Tucker GT (1996) 1-Methyl-4-phenyl-1,2,3,6-tetahydropyridine (MPTP) is N-demethylated by cytochromes P450 2D6, 1A2 and 3A4 – implications for susceptibility to Parkinson's disease. J Pharmacol Exp Ther 277:685–690

Daly AK, Brockmuller J, Broly F, Eichelbaum M, Evans WE, Gonzalez FJ, Huang JD, Idle JR, Ingelmansundberg M, Ishizaki T, Jacqzaigrain E, Meyer UA, Nebert DW, Steen VM, Wolf CR, Zanger UM (1996) Nomenclature for human CYP2D6 alleles. Pharmacogenetics 6:193–201

Doll R, Peto R (1981) The causes of cancer: Quantative estimates of avoidable risks of cancer in the United States today. J Natl Cancer Inst 66:1192–1308

Downes CP, Wolf CR, Lane DP (eds) (1998) Biochemical symposium no. 64: Cellular responses to stress. Portland Press Ltd, London

Forrester LM, Henderson CJ, Glancey MJ, Back DJ, Park BK, Ball SE, Kitteringham NR, McLaaren AW, Miles JS, Skett P, Wolf CR (1992) Relative expression of cytochrome P450 isozymes in human liver and association with the metabolism of drugs and xenobiotics. Biochem J 281:359–368

Gibson GG, Skett P (1994) Introduction to drug metabolism. Blackie Academic and Professional, Glasgow

Gilham DE, Cairns W, Paine MJI, Modi S, Poulsom R, Roberts GCK, Wolf CR (1997) Metabolism of MPTP by cytochrome P4502D6 and the demonstration of 2D6 mRNA in human foetal and adult brain by in situ hybridisation. Xenobiotica 27:111–125

Gonzalez FJ, Nebert DW (1990) Evolution of the P450 gene superfamily: animal-plant "warfare", molecular drive and human genetic differences in drug oxidation. Trends Genet 6:182–186

Gonzalez FJ, Skoda RC, Kimura S, Umeno M, Zanger UM, Nebert DW, Gelboin HV, Hardwick JP, Meyer UA (1988) Characterisation of the common genetic defect in humans deficient in debrisoquine metabolism. Nature 33:442–446

Gough AC, Miles JS, Spurr NK, Moss JE, Gaedigk A, Eichelbaum M, Wolf CR (1990) Identification of the primary gene defect at the cytochrome P450 CYP2D locus. Nature 347:773–776

Hayes JD, Wolf CR (1990) Molecular mechanisms of drug resistance. Biochem J 272:281–295

Hayes JD, Pulford DJ (1995) The glutathione S-transferase supergene family: regulation of GST and the contribution of the isozymes to cancer chemoprotection and drug resistance. Crit Rev Biochem Mol Biol 30:445–600

Hayes JD, Wolf CR (eds) (1997) Molecular genetics of drug resistance. Harwood Academic Publishers, London

Henderson CJ, Smith AG, Ure J, Brown K, Bacon EJ, Wolf CR (1998) Increased skin tumorigenesis in mice lacking pi class glutathione S-transferases. Proc Natl Acad Sci USA 95:5275–5280

Islam SA, Wolf CR, Lennard MS, Sternberg MJE (1991) A three-dimensional molecular template for substrates of human cytochrome P450 involved in debrisoquine 4-hydroxylation. Carcinogenesis 12:2211–2219

Issemann I, Green S (1990) Activation of a member of the steroid hormone receptor superfamily by peroxisome proliferators. Nature 347:645–649

Jarpe MB, Widmann C, Knall C, Schlesinger TK, Gibson S, Yujiri T, Fanger GR, Gelfand EW, Johnson GL (1998) Anti-apoptotic versus pro-apoptotic signal transduction: Checkpoints and stop signs along the road to death. Oncogene 17:1475–1482

Johansson I. Lundqvist E, Bertilsson L, Dahl ML, Sjoqvist F, Ingelmansundberg M (1993) Inherited amplifications of an active gene in the cytochrome P450 CYP2D locus as a cause of ultrarapid metabolism of debrisoquine. Proc Natl Acad Sci USA 90:11825–11829

Kalow W (ed) (1992) Pharmacogenetics of drug metabolism. Pergamon Press, New York

Kerppola TK, Curran T (1994) Maf and Nrl can bind to AP1 sites and form heterodimers with Fos and Jun. Oncogene 9:675–684

Lehmann JM, McKee DD, Watson MA, Wilson TM, Moore JT, Kliewer SA (1998) The human orphan nuclear receptor pxr is activated by compounds that regulate CYP3A4 gene expression and cause drug interactions. J Clin Invest 102:1016–1023

Lown KS, Mayo RR, Leichtman AB, Hsiao H, Turgeon K, Schmiedlin-Ren P, Brown MB, Guo W, Rossi SJ, Benet LZ, Watkins PB (1997) Pharmacokinetics and drug disposition: Role of intestinal P-glycoprotein (mdr1) in interpatient variation in the oral bioactivity of cyclosporine. Clin Pharmacol Ther 62:248–260

Nebert DW, Puga A, Vasiliou V (1993) Role of the Ah receptor and the dioxin-inducible (Ah) gene battery in toxicity, cancer and the signal transduction. Ann NY Acad Sci 685:624–640

Nelson DR, Koymans L, Kamataki T, Stegeman JJ, Feyereisen R, Waxman DJ, Waterman MR, Gotoh O, Coon MJ, Estrabrook RW, Gunsalus IC, Nebert DW (1996) P450 superfamily: update on new sequences, gene mapping, accession numbers and nomenclature. Pharmacogenetics 6:1–42

Potter JD (ed) (1997) Food, nutrition and the prevention of cancer: A global perspective. World Cancer Research Fund/American Institute for Cancer Research, Washington

Reisz-Porszasz S, Probst MR, Fukunaga BN, Hankinson O (1994) Identification of functional domains of the aryl hydrocarbon receptor nuclear translocator protein (ARNT). Mol Cell Biol 14:6075–6086

Rushmore TH, King RG, Paulson KE, Pickett CB (1990) Regulation of glutathione S-transferase Ya subunit gene expression – Identification of a unique xenobiotic-responsive element controlling inducible expression by planar aromatic-compounds. Proc Natl Acad Sci USA 87:3826–3830

Ryberg D, Skaug V, Hewer A, Phillips DH, Harries LW, Wolf CR (1997) Genotypes of glutathione transferase M1 and P1 and their significance for lung DNA adduct levels and cancer risk. Carcinogenesis 18:1285–1289

Shimada T, Yamazaki H, Mimura M, Inui Y, Guengerich FP (1994) Inter-individual variations in human liver cytochrome P450 enzymes involved in the oxidation of drugs, carcinogens and toxic chemicals: studies with liver microsomes of 30 Japanese and 30 Caucasians. J Pharmacol Exp Ther 270:414–423

Smith CAD, Gough AC, Leigh PN, Summers BA, Harding AE, Maranganore DM, Sturman SG, Schapira AHV, Williams AC, Spurr NK, Wolf CR (1992) Debrisoquine hydroxylase gene polymorphism and susceptibility to Parkinson's disease. Lancet 339:1375–1377

Smith G, Stanley LA, Sim E, Strange RC, Wolf CR (1995) Metabolic polymorphisms and cancer susceptibility. Cancer Surveys 25:27–65

Sueyoshi T, Kawamoto T, Zelko I, Honkakoski P, Negishi M (1999) The repressed nuclear receptor CAR responds to phenobarbital in activating the human CYP2B6 gene. J Biol Chem 274:6043–6046

Watkins PB (1997) The barrier function of CYP3A4 and P-glycoprotein in the small bowel. Adv Drug Deliv Rev 27:161–170

Williams RT (1959) Detoxification mechanisms. Chapman and Hall, London

Wolf CR (1986) Cytochrome P450s: Polymorphic multigene families involved in carcinogen activation. Trends Genet 2:209–214

Wolf CR (1990) Metabolic factors in cancer susceptibility. Cancer Surveys 9:437–474

Wolf CR, Smith G (1999) Pharmacogenetics. Br Med Bull 55:366–386

Wolff T, Distlerath LM, Worthingham MT, Groopman ID, Hammons GJ, Kadlubar FF, Prough RA, Martin MV, Guengerich FP (1985) Substrate specificity of human liver cytochrome P450 debrisoquine 4-hydroxylase probed using immunochemical inhibition and chemical modeling. Cancer Res 45:2116–2122

Leptin and the Neural Circuit Regulating Body Weight

J. M. Friedman

Leptin, the hormone encoded by the ob gene, plays an important role in regulating the size of the adipose tissue mass. Injections of recombinant leptin reduce body weight and food intake of normal and obese (ob) mice in a dose dependent manner but have no effect on diabetic (db) mice, another recessive obesity gene. These data identify leptin as an important signaling molecule that acts to maintain constant stores of body fat. The complete insensitivity of db mice to leptin and the identical phenotype of ob and db mice suggested that the db locus encodes the leptin receptor. The db gene was found to be identical to a leptin receptor (Ob-R) that was functionally cloned from choroid plexus. However, because this receptor was normal in C57BL/Ks db/db mice, the possibility was raised that this db mutation affected an alternatively spliced form. The Ob-R gene was found to encode at least five different splice variants. One of the splice variants is expressed at a high level in the hypothalamus and at a lower level in other tissues. This transcript is mutant in C57BL/Ks db/db mice. The mutant protein is missing the cytoplasmic region and is defective in signal transduction. Further studies have revealed that the STAT3 transcription factor is activated specifically in hypothalamus within 15 minutes of a single injection of leptin in ob and wild type mice but not in db mice. In vitro studies indicate that SHP-2, a phosphoprotein phosphatase, is also a component of the leptin signal transduction pathway. Consistent with a CNS site of action, low-dose infusions of leptin (5 ng/hr) intracerebroventicularly reproduce the effects of much larger doses given peripherally. To elucidate the components of the neural circuit activated by leptin, novel methods for neural tracing are being developed and will be discussed.

The nature of the efferent signals from integratory centers in the hypothalamus is also under study. Infusions of leptin ICV result in changes in fat and glucose metabolism that are qualitatively different from those that result from food restriction. Studies of the metabolic response to leptin and the mechanisms by which CNS signals affect glucose and fat metabolism are also underway.

Cloning of the *ob* Gene and Characterization of Leptin

Recessive mutations in the mouse *ob* and *db* genes result in obesity and diabetes in a syndrome resembling morbid human obesity (Friedman and Halaas 1998). Affected *ob* and *db* mice have identical phenotypes, with each mutant weighing

V. Boulyjenkov, K. Berg, Y. Christen (Eds.)
Genes and Resistance to Diseases
© Springer-Verlag Berlin Heidelberg 2000

three times as much as normal mice, with a five-fold increase in body fat content. Previous data suggested that the *ob* gene encoded a novel hormone that regulated weight and that the *db* gene encoded the receptor for this hormone (Coleman and Hummel 1970). Confirmation of this hypothesis, put forth by Doug Coleman, awaited the identification of the *ob* and *db* genes. Positional cloning was used to identity the *ob* and *db* genes (Zhang et al. 1994; Halaas et al. 1995; Lee et al. 1996).

The *ob* gene encodes a novel 167 amino acid protein, now known as leptin. This protein sequence is novel and has no homologues in the public sequence database. Leptin is not produced in the two available ob alleles. C57B1/6J *ob/ob* mice have a nonsense mutation at codon 105 (Zhang et al. 1994). 129 *ob2J/ob2J* mice carry a transposon insertion that ablates expression of ob RNA (Moon and Friedman 1997). Leptin has a functional signal sequence, which initially suggested that it was secreted. In addition, the level of its RNA was increased in *ob* and *db* mice, indicating that the *ob* gene is under feedback control (Zhang et al. 1994; Maffei et al. 1995a).

These data suggested the hypothesis that leptin is an afferent signal in a negative feedback loop regulating the size of adipose tissue mass. However, this theory was unproven and required further experiments. To establish that leptin functions as a hormone controlling body fat content, the following criteria had to be satisfied: 1) the *ob* gene should be expressed in adipocytes, the principal site of fat storage; 2) it should circulate in plasma; 3) plasma leptin levels should be increased after weight gain and decreased with weight loss; and 4) recombinant protein should reduce body fat content when injected into ob and wild type mice but not *db* mice.

In situ hybridization and cell fractionation were used to demonstrate that ob RNA is expressed only in adipocytes (Maffei et al. 1995a). Leptin circulates as a 16 kilodalton protein mouse and human plasma but is undetectable in plasma from C57BL/6J *ob/ob* mice (Halaas et al. 1995). Circulating leptin is not post-translationally modified (Cohen et al. 1996). Plasma levels of this protein are increased in *db* mice, a mutant since confirmed to be resistant to the effects of leptin. The levels of protein are also increased in several other genetic and environmentally induced forms of rodent obesity including mice with lesions in the hypothalamus (Maffei et al. 1995b). Finally, the plasma levels of leptin fall in both humans and mice after weight loss (Maffei et al. 1995b).

Administration of recombinant leptin leads to a decrease of body weight in mice and other mammals. Daily intraperitoneal injections of recombinant mouse leptin reduced body weight of *ob/ob* mice by 30 % at two weeks and by 40 % after four weeks but had no effect on *db/db* mice (Halaas et al. 1995). The protein reduced food intake and increased energy expenditure in *ob/ob* mice. Twice daily of mouse protein into wild type mice resulted in a sustained 12 % weight loss, decreased food intake and a reduction of body fat from 12.2 to 0.7 %. In contrast to diet-induced weight loss, leptin spared lean body mass. Initially this effect required relatively high doses. However, the dosage requirements are greatly reduced when the protein is administered by constant subcutaneous infusion.

A physiologic increase in plasma leptin reduces weight in a dose-dependent manner and plasma levels of recombinant leptin of 25 ng/ml (5 ng/ml is the normal level of leptin in mouse plasma) are sufficient to totally deplete adipose tissue mass (Halaas et al. 1997). Conversely, decreasing leptin levels signal nutrient deprivation and lead to a compensatory response that includes positive energy balance and other hormonal and metabolic immunologic changes (Friedman and Halaas 1998). These data confirm that leptin serves an endocrine function to regulate body fat stores.

Molecular Cloning of the *db* Gene: Studies of Leptin's Site of Action

The complete insensitivity of *db* mice to leptin and the identical phenotype of *ob* and *db* mice suggested that the *db* locus encoded the leptin receptor (Halaas et al. 1995; Coleman and Hummel 1970; Maffei et al. 1995a). Additional experiments on mice with lesions of the hypothalamus led to the conclusion that leptin's effects were mediated via effects on the hypothalamus, a brain region known to control body weight (Maffei et al. 1995a). Confirmation of this required the identification of the leptin receptor and/or the *db* mutation. The leptin receptor (initially cloned by Tartaglia et al.) was found to be encoded by the *db* gene (Lee et al. 1996; Tartaglia et al. 1995). The gene encodes at least five alternatively spliced forms. Four of the splice forms encode transmembrane proteins whereas one, Ob-Re, encodes a circulating form of the leptin receptor (Lee et al. 1996; Li et al. 1998). One of the splice variants, Ob-Rb, is expressed at a high level in the hypothalamus and at a lower level in other tissues. This transcript is detective specifically in C57BL/Ks db/db mice, one of the mutant db strains (Lee et al. 1996; Tartaglia et al. 1995). The mutation is the result of abnormal splicing leading to a 106 bp insertion into the 3' end of its RNA. The mutant protein is missing the cytoplasmic region and is likely to be detective in signal transduction (Lee et al. 1996; Vaisse et al. 1996). A nonsense mutation in facp rats, a rat equivalent of *db*, leads to premature termination of the NH2-terminal of the transmembrane domain (Lee et al. 1997). In *db3J* mice, a frame shift mutation in the extracellular region ablates expression of the wild type receptor protein (Lee et al. 1997). In *db*[Pas] mice a duplication also leads to premature termination of leptin receptor synthesis (Li et al. 1998). The phenotypes of C57BL/6J *db/db* mice (mutant in Ob-Rb), *db*[Pas] and 129 *db3J/db3J* mice (mutant in all forms) are identical. These data suggest that the weight-reducing effects of leptin are mediated by activation of signal transduction by Ob-Rb in the hypothalamus and perhaps elsewhere.

Further evidence in support of a hypothalamic site of action was generated using in vivo assays of the STAT transcription factors in response to leptin (Vaisse et al. 1996). Ob-R is a member of the cytokine family of receptors. These receptors regulate transcript ion via tyrosine phosphorylation of members of the STAT family of transcript ion factors. STAT activation was assayed by gel shift using nuclear extracts from a number of mouse tissues after leptin treatment. Dose-dependent activation of STAT3 is demonstrable in the hypothalamus of mice within 15 minutes of a single intravenous injection of leptin (Vaisse et al. 1996). This effect is not seen in C57B1/KS *db/db* mice, which have a defect in the Ob-Rb isoform of the leptin receptor. Additionally, Ob-Rb does not lead to the activation of any known STATs in response to leptin in any other tissues. SKP-2, a phosphotyrosine phosphatase, is also activated by leptin (Li and Friedman, submitted for publication). Neurons in hypothalamus that express Ob-Rb have been identified in the arcuate, VMH and LH nuclei (Fei et al. 1997). These nuclei have all been shown to play a role in regulating food intake and weight. Leptin acts on neurons in these nuclei and activates signal transduction by phosphory-lating STAT3, SHP-2 and other proteins (Li and Friedman, submitted).

The effect of peripherally administered leptin has been compared to that of leptin administered directly into the mouse III ventricle using Alzet osmotic infusion pumps (ICV; Halaas et al. 1997). Infusion of leptin ICV at a dose of 3 ng/hour results in a more profound response (20 % weight loss at seven days) than that seen in response to a twice-daily intraperitoneal dose of 250 ug (12 % weight loss). The effect of the 3 ng/hour ICV dose appears to be at the peak of the dose response curve, as higher doses (50 and 500 ng/hour) do not have a greater effect. To date, all of the effects of peripheral leptin tested thus far are reproduced by leptin infused into the III ventricle. These data establish leptin as the most potent weight-reducing peptide when delivered into the CSF and strongly support the hypothesis that the CNS is a major site of leptin action. While these data suggest a CNS site of action, they do not exclude the possibility that leptin also acts on other tissues.

Novel efferent signals from the hypothalamus are activated by leptin. Infusions of leptin result in changes in fat and glucose metabolism that are qualitatively different from those that follow food restriction and lead to the disappearance of adipose tissue mass without decreasing lean body mass (Halaas et al. 1995). Leptin also stimulates glucose uptake by skeletal muscle and brown adipose tissue (Kamohara et al. 1997). These data have suggested that leptin may improve diabetes independent of its effects on weight, a possibility now being explored in clinical trials (Halaas et al. 1995; Kamohara et al. 1997).

Relevance of Leptin to Other Forms of Obesity

The relationship of leptin to other forms of obesity can be interred by measurement of plasma leptin level. An increase in plasma levels suggests that obesity is the result of a downstream effect leading to leptin resistance. A low or normal plasma concentration of leptin in the context of obesity suggests that there is decreased leptin synthesis and/or secretion. This postulate is similar to that which relates insulin to the pathogenesis of diabetes. An ELISA has been used to measure plasma [leptin] in a number of animal obesities and in human. In humans ~90 % of obese subjects have elevated leptin levels whereas 10 % have normal levels (Maffei et al. 1995b). In all forms of rodent obesity studied to date, the obese animals have a higher leptin level than controls (not including *ob* mice and *ob/ob* transgenic mice, described below (Maffei et al. 1995b). The animals studied included mutant Ay, fat and tub mice, diet-induced obese mice, New Zealand Obese mice (NZO), old mice and mice with hypothalamic lesions. These data suggest that these forms of animal obesity appear to be the result of leptin resistance. Indeed, the sensitivity of Ay, NZO and DIO mice to intraperitoneal leptin is substantially reduced (Halaas et al. 1997). However, the mechanism of leptin resistance is different in each of these three cases. DIO mice are partially resistant to leptin, Ay mice appear to have a detect in the neural circuit activated by leptin and NZO mice have a detect in leptin transport into the CNS.

The fact that some obese subjects have normal (inappropriately low) leptin levels suggests that decreased production can also lead to obesity. This possibility is supported by the observation that abnormally low expression of the *ob* gene leads to moderately severe obesity in mice (Ioffe et al. 1998). Taken together, these results suggest that the pathogenesis of leptin resistance and obesity is likely to be complex and analogous to insulin resistance in Type II diabetes.

Leptin in Human Physiology

As mentioned, plasma leptin concentration correlates with body fat content and is usually increased in obese subjects (Maffei et al. 1995b; Considine et al. 1996). This finding suggests that human obesity is generally associated with insensitivity to leptin. However, 5–10 % of obese human subjects have relatively low levels of leptin, suggesting a reduced rate of leptin production in this subgroup (Maffei et al. 1995b; Considine et al. 1996). Low leptin levels also predispose pre-obese Pima Indians to weight gain (Ravussin et al. 1997).

In humans, diet-induced weight loss results in a decrease in plasma leptin concentration (Maffei et al. 1995b; Considine et al. 1996). This may explain the high failure rate of dieting, as low leptin is likely to be a potent stimulus to weight gain. Anorexia nervosa patients also have been reported by others to have extremely low leptin levels (Casanueva et al. 1997; Mantzoros et al. 1997). Refeeding of these patients results in a rapid increase in plasma leptin concentration to roughly normal levels before normal weight is achieved. Thus, excessive leptin production could play a permissive role in the pathogenesis of this condition.

In almost all cases, obese subjects express at least some leptin, indicating that human *ob* mutations are likely to be rare (Maffei et al. 1995b). Indeed there were no *ob* mutations in one study in which ~ 500 obese subjects were tested (Maffei et al. 1996). Nevertheless, a few *ob* and *db* mutations have been described by other investigators. Two cousins homozygous for a frameshift mutation in the leptin gene are markedly obese and do not have any circulating leptin (Montague et al. 1997). Three members of a Turkish kindred with a missense mutation in the leptin gene are extremely obese and amennorhoeic, indicating that leptin is important in modulating human reproductive function (Strokel et al. 1998). Three massively obese members of a French family carry mutations in the leptin receptor and have reproductive abnormalities (Clement et al. 1998).

The association of mutations in leptin and its receptor with massive obesity confirms its importance in regulating human body weight. However, these syndromes are rare. The pathogenesis of most human obesity is still unknown and likely to be the result of differences in leptin secretion and/or leptin sensitivity. However, this still leaves open the possibility that leptin may prove useful in the treatment of obesity and diabetes. Clinical trials of leptin are now underway at a number of sites. The available data suggest that a subset of obese humans may in fact lose weight on leptin therapy. Further studies are required to confirm this possibility. Whether or not leptin emerges as a human therapeutic agent, a fuller understanding of the molecular mechanisms that regulate both leptin production and leptin's effects on hypothalamic neurons and other cell types will have important implications for understanding the pathogenesis of obesity and other nutritional disorders. Studies of leptin are also likely to reveal additional links between nutritional state and animal physiology.

References

Casanueva FF, Dieguez C, Popovic V, Peino R, Considine RV, Caro JF (1997) Serum immunoreactive leptin concentrations in patients with anorexia nervosa before and after partial weight recovery. Biochem Mol Med 60:116–120

Clement K, Vaisse C, Lahlou N, Cabrol S, Pelloux V, Cassuto D, Gourmelen M, Dina C, Chambaz J, Lacorte JM, Basdevant A, Bougneres P, Lebouc Y, Froguel P, Guy-Grand B (1998) A mutation in the human leptin receptor gene causes obesity and pituitary dysfunction. Nature 392(6674):398–401

Cohen SL, Halaas JL, Friedman JM, Chait BT, Bennett L, Chang D, Hecht R, Collins F (1996) Human leptin characterization [letter]. Nature 382:589

Coleman DL, Hummel KP (1970) The effects of hypothalamic lesions in genetically diabetic mice. Diabetologia 6:263–267

Considine RV, Sinha MK, Heiman ML, Kriauciunas A, Stephens TW, Nyce MR, Ohannesian JP, Marco CC, McKee LJ, Bauer TL, Caro JF (1996) Serum immunoreactive-leptin concentrations in normal-weight and obese humans New Engl J Med 334:324–325

Fei H, Okano HJ, Li C, Lee G-H, Zhao C, Darnell R, Friedman JM (1997) Anatomic localization of alternatively spliced leptin receptors (Ob-R) in mouse brain and other tissues. Proc Natl Acad Sci USA. 94:7001–7005

Friedman J, Halaas J (1998) Leptin and the regulation of body weight in mammals. Nature 395:763–770

Halaas JL, Gajiwala KS, Maffei M, Cohen SL, Chait BT, Rabinowitz D, Lallone RL, Burley SK, Friedman JM (1995) Weight-reducing effects of the plasma protein encoded by the obese gene. Science 269:543–546.

Halaas JL, Boozer C, Blair-West J, Fidahusein N, Denton D, Friedman JM (1997) Physiological response to long-term peripheral and central leptin infusion in lean and obese mice. Proc Natl Acad Sci USA 94:8878–8883

Ioffe E, Moon B, Connolly E, Friedman JM (1998) Abnormal regulation of the leptin gene in the pathogenesis of obesity. Proc Natl Acad Sci USA 95:11852–11857.

Kamohara S, Burcelin R, Halaas JL, Friedman JM, Charron MJ (1997) Acute intravenous and intracerebroventricular leptin infusion increases glucose uptake and glucose turnover by an insulin independent mechanism. Nature 389:374–377

Lee GH, Proenca R, Montez JM, Carroll KM, Darvishzadeh JG, Lee JI, Friedman J (1996) Abnormal splicing of the leptin receptor in diabetic mice. Nature 379:632–635

Lee G-H, Li C, Montez J, Halaas J, Darvishzadeh J, Friedman JM (1997) Leptin receptor mutations in 129 db3J/db3J mice and NIH facp/facp rats. Mammalian Genome 8:445–447

Li C, Ioffe E, Fidahusein N, Connolly E, Friedman JM (1998) Absence of soluable leptin receptor in plasma from dbPas/dbPas and other db/db mice. J Biol Chem 273:10078–10082

Maffei M, Fei H, Lee GW, Dani C, Leroy P, Zhang Y, Proenca R, Negrel R, Ailhand G, Friedman JM (1995a) Increased expression in adipocytes of ob RNA in mice with lesions of the hypothalamus and with mutations at the db locus. Proc Natl Acad Sci USA 92:6957–6960

Maffei M, Halaas J, Ravussin E, Pratley RE, Lee GH, Zhang Y, Fei H, Kim S, Lallone R, Ranganathan S, Kern PA, Friedman JM (1995b) Leptin levels in human and rodent: Measurement of plasma leptin and ob RNA in obese and weight-reduced subjects. Nature Med 1:1155–1161

Maffei M, Stoffel M, Barone M, Moon B, Dammerman M, Ravussin E, Bogardus C, Ludwig DS, Flier JS, Talley M, Auerbach S, Friedman JM (1996) Absence of mutations in the human ob gene in obese/diabetic subjects. Diabetes 45:679–682

Mantzoros C, Flier JS, Lesem MD, Breweton TD, Jimerson DC (1997) Cerebrospinal fluid leptin in anorexia nervosa: correlation with nutritional status and potential role in resistance to weight gain. J Clin Endocrinol Metab 82:1845–1851

Montague CI, Farooqi IS, Whitehead JP, Soos MA, Rau H, Wareham NJ, Sewter CP, Digby JE, Mohammed SN, Hurst JA, Cheetham CH, Earley AR, Barnett AH, Prins JB, O'Rahilly S (1997) Congenital leptin deficiency is associated with severe early-onset obesity in humans. Nature 387:903–908

Moon BC, Friedman JM (1997) The molecular basis of the obese mutation in ob2J mice. Genomics 42:152–156

Ravussin E, Pratley RE, Maffei M, Wang H, Friedman JM, Bennett PH, Bogardus C (1997) Relatively low plasma leptin concentrations precede weight gain in Pima Indians. Nature Med 3:238–240

Strobel A, Camoin TIL, Ozata M, Strosberg AD (1998) A leptin missense mutation associated with hypogonadism and morbid obesity. Nature Genet 18:213–215

Tartaglia LA, Dembski M, Weng X, Deng N, Culpepper J, Devos R, Richards GF, Campfield LA, Clark FT, Deeds J, Muir C, Sanker S, Moriarty A, Moore KJ, Smutko JS, Mays GG, Woolf EA, Monroe CA, Tepper RI (1995) Identification and expression cloning of a leptin receptor, OB-R Cell 83:1263–1271

Vaisse C, Halaas JL, Horvath CM, Darnell JE, Jr., Stoffel M, Friedman JM (1996) Leptin activation of Stat3 in the hypothalamus of wild-type and ob/ob mice but not db/db mice. Nature Genet 14:95–97

Zhang Y, Proenca P, Maffei M, Barone M, Leopold L, Friedman JM (1994) Positional cloning of the mouse obese gene and its human homologue. Nature 372:425–432

The Cholesteryl Ester Transfer Protein (CETP) Locus and Protection Against Atherosclerosis

K. Berg

Atherosclerosis and Risk Factors: Genes and Environment

Atherosclerotic diseases, including coronary heart disease (CHD), were long considered to be the results of unhealthy diet and life style. Heart attacks in people with autosomal dominant familial hypercholesterolemia (FH) were considered rather exceptional cases where genes caused CHD. Only over the last two or three decades has it become widely accepted that genetic as well as nutritional or life style factors and their interaction are causative with respect to CHD. Studies focusing on overt disease conducted in the 1970s uncovered a significant aggregation of (particularly younger) cases of CHD in families which could not be explained only by shared environment or diet. It became clear that having a close relative who contracted CHD at a relatively young age was a risk factor in its own right (Nora et al. 1980). One could not avoid the conclusion that genes contributed to disease risk (for review, see Berg 1983, 1993; Nora et al. 1991).

Biological as well as life style factors associated with CHD risk were identified; cigarette smoking, hypertension and elevated cholesterol level are the CHD risk factors that have been most thoroughly examined. These risk factors were thought of as primarily environmental in nature, with stress or unhealthy life style being blamed for hypertension, and a too-high consumption of fats and calories being blamed for elevated cholesterol levels. However, studies of twins and other relatives showed that several biochemical variables that are risk factors or protective factors with respect to CHD are significantly influenced by genes (for review, see Berg 1983, 1993; Nora et al. 1991). In our own twin studies, we have observed high levels of heritability for total cholesterol (0.68), apolipoprotein B (apoB) (0.64), high density lipoprotein (HDL) cholesterol (HDLC) (0.71), apolipoprotein A-I (apoA-I) (0.55), and apolipoprotein A-II (apoA-II) (0.68). The Lp(a) lipoprotein level exhibits heritability very close to unity.

The high heritability of components of HDL is of particular interest in the present context. It has been re-examined and confirmed in several models, and significant heritability levels for apoB, homocystein, blood pressure and body mass index have also been confirmed in new series.

V. Boulyjenkov, K. Berg, Y. Christen (Eds.)
Genes and Resistance to Diseases
© Springer-Verlag Berlin Heidelberg 2000

Searching for the Genes

The earliest genetic studies on total cholesterol were conducted with random genetic markers such as blood groups and some interesting associations were uncovered. With developments in the area of DNA technology, it has become possible to conduct much more targeted genetic analyses, as well as total genome searches.

In recent years, numerous studies following a functional candidate gene approach have been reported. With respect to atherosclerotic disease, a functional candidate gene is any gene expressed in atherosclerotic lesions, or whose protein product is, or could be, involved in
- lipoprotein structure or metabolism
- thrombogenesis, thrombolysis or fibrinolysis
- regulation of blood flow in coronary arteries
- regulation of blood pressure
- regulation of growth of atherosclerotic lesions
- early development of coronary arteries
- reverse cholesterol transport

This paper will focus on the last item, reverse cholesterol transport.

Reverse Cholesterol Transport and Cholesteryl Ester Transfer Protein

For the body to get rid of cholesterol from the tissues, cholesterol must be transported to the liver to be excreted in the bile. It is the transportation of cholesterol from peripheral cells to the liver that is referred to as reverse cholesterol transport. Variations in the efficiency of removing excess cholesterol from peripheral tissues, including the vasculature, could be essential in atherogenesis.

Reverse cholesterol transport (Tall 1993) starts with small, cholesterol-poor, pre-β migrating HDL, containing apoA-I, picking up cholesterol from peripheral cells. Cholesterol taken up by HDL is esterified by lecithin-cholesterol acyl transferase and HDL matures to spherical particles. Cholesteryl esters formed in HDL are transferred to apoB-containing lipoproteins through the action of cholesteryl ester transfer protein (CETP) and transported to the liver to be excreted in the bile. CETP as well as hepatic lipase promote dissociation of apoA-I from mature HDL, to generate pre-β HDL particles that could accelerate the transfer of cellular cholesterol to HDL. Triglycerides entering HDL in exchange for cholesteryl esters are hydrolyzed on HDL by hepatic lipase (Fielding and Havel 1996).

Taking part in the process of moving excess cholesterol from peripheral cells to the liver would seem to make CETP anti-atherogenic. However, CETP-deficient people of Japanese origin did not show clinical signs of advanced atherosclerosis in early reports (Tall 1993). The increase in cholesteryl esters in very low density lipoprotein (VLDL) and low density lipoprotein (LDL) promoted by CETP could, in fact, increase the atherogenic capacity of circulating VLDL and LDL. This has led to the belief that CETP may be "pro-atherogenic." It has been found that transgenic mice expressing high levels of the simian CETP gene are more suscep-tible to an atherogenic, fat-rich diet than normal mice (Marotti et al. 1993). Thus, there is not a simple relationship between CETP, HDL and CHD risk.

Zhong et al. (1996) studied Japanese-American men with mutations causing partial CETP deficiency. At high or low HDLC concentrations, CETP deficiency was without effect but at intermediate values of HDLC, CETP deficiency was associated with an increased incidence of CHD (Zhong et al. 1996). When CETP-deficient patients with very high HDLC concentrations have CHD, they may also have low hepatic lipase activity and, therefore, a deficiency of both mechanisms for generating lipid-poor apoA-I-containing particles (Fielding and Havel 1996).

Although CETP activity and HDLC appear to be negatively correlated (Han-nuksela et al. 1994), the relationship between the two variables is complex. The role of CETP may depend on the metabolic context (Kuivenhoven et al. 1998) and CETP as well as HDL are affected by smoking and to some extent by alcohol con-sumption.

The CETP Gene and its Normal Variants

Drayna et al. (1987) cloned and sequenced cDNA representing the CETP gene. The gene is located on chromosome 16 (16q12–21; Lusis et al. 1987) and spans about 25 kilobases (kb). It has 16 exons. Drayna and Lawn (1987) detected nor-mal polymorphisms demonstrable with the restriction enzyme TaqI in the CETP gene and several additional polymorphisms have since been discovered. Of the CETP polymorphisms, the "B" polymorphism detectable with the restriction enzyme TaqI (the "TaqIB" polymorphism) has been particularly well studied. This polymorphism resides in intron 1 of the CETP gene and is therefore not causing any change in CETP structure. The rationale for studying the polymor-phism with respect to biochemical function or clinical relevance is that it could be in linkage disequilibrium with functionally important domains in or near the CETP gene.

The CETP-TaqIB Polymorphism is associated with Clinically Relevant HDL Variables

In our search for genes contributing to the population variation in CHD risk fac-tors, we examined the TaqIB polymorphism and the levels of apoA-I, apoB, total serum cholesterol, HDLC, triglycerides and Lp (a) lipoprotein and reported our findings in 1989 (Kondo et al. 1989). Among healthy people, we found signifi-cantly higher levels of HDLC (p = 0.03) and apoA-I (p = 0.005) in persons who were homozygous for absence of the restriction site (genotype B2B2) than in people who were homozygous for presence of the site (genotype B1B1). Hetero-zygotes had intermediate levels. There was no significant difference between genotypes with respect to levels of total cholesterol, triglycerides, apoB or Lp (a) lipoprotein.

Table 1. Sex and age adjusted mean HDLC and apoA-I levels in smokers (n = 59) and non-smokers (n = 86) according to genotype in the TaqIB polymorphism at the CETP locus. (Extracted from Kondo et al. 1989)

Genotype	HDLC (mmol/l) Smokers	Non-smokers	ApoA-I (mg/dl) Smokers	Non-smokers
B1B1	1.18	1.34	141	138
B1B2	1.27	1.48	141	152[a]
B2B2	1.32	1.54	147	162[b]

[a] Significance of difference from B1B1 group: p = 0.017 (for measured as well as log 10 transformed values)

[b] Significance of difference from B1B1 group: p = 0.007 (for measured values, p = 0.006 for log 10 transformed values)

Smoking obliterates a beneficial Gene Effect

Since smoking is known to affect the levels of HDL cholesterol and apoA-I, the analyses were also conducted separately for smokers and non-smokers. In non-smokers, both B1B2 heterozygotes and B2B2 homozygotes had higher levels of apoA-I than B1B1 homozygotes and the difference between the two categories of homozygotes was highly significant (p = 0.007). In smokers, B1B1 homozygotes and B1B2 heterozygotes had the same apoA-I level and this was similar to the level in non-smoking B1B1 homozygotes (Table 1). The slightly higher apoA-I level in smoking B2B2 homozygotes than in the other genotypic groups of smokers did not reach statistical significance.

   Thus, we had uncovered an association between normal genetic variation at a candidate locus with respect to CHD risk and a biochemical variable relevant to CHD risk that was obliterated in smokers. High levels of HDL are protective with respect to CHD, and the study showed that smoking prevented people with CETP genotype B1B2 or B2B2 from sustaining a HDL concentration at the level that their genetic constitution would permit them to achieve. It appeared that we had identified a sub-group of people whose genes might protect them against atherosclerotic disease, provided that they abstained from smoking.

Association between B2 Allele and High HDL Level Confirmed

The association between high HDL level and the B2 gene of the TaqIB polymorphism may well be the one association between a genetic marker and a CHD risk or protective factor that has been most consistently confirmed (Freeman et al. 1990; Mendis et al. 1990; Hannuksela et al. 1994; Mitchell et al. 1994; Fumeron et al. 1995; Kauma et al. 1996; Dullaart et al. 1997, 1998; Kuivenhoven et al. 1997; Song et al. 1997; Bernard et al. 1998). The association has been confirmed in different ethnic groups, in healthy people and in patient populations (including diabetics). The effect of smoking was also confirmed at an early stage and some workers have found that alcohol intake may modulate the effect of this polymorphism on HDL (Fumeron et al. 1995, Hannuksela et al. 1994) whereas others have seen no effect of moderate alcohol consumption (Dullaart et al. 1998). In some studies there have been differences between sexes with respect to the effect of this polymorphism on HDL (Kauma et al. 1996), and other studies have suggested an effect of obesity (Freeman et al. 1994).

The TagIB Polymorphism, CETP Level and HDL Level

The cholesteryl ester transfer activity appears to be higher in B1B1 homozygotes than in B2B2 homozygotes, in agreement with an observed trend towards negative correlation between CETP activity and HDL concentration. However, the effects of the CETP polymorphism on CETP and HDL levels have been reported to be independent, suggesting the presence of at least two functional variants linked to B1 (Fumeron et al. 1995). Multivariate analysis has indicated that the TaqIB polymorphism has an effect on HDL that is independent of age, sex, body mass index, oral contraceptive use, exercise, alcohol consumption and plasma triglycerides (Freeman et al. 1994).

Effect of Other Polymorphisms in the CEPT Gene

Several polymorphisms other than the TaqIB polymorphism in the CETP gene have been examined (Arai et al. 1996; Bruce et al. 1998; Bu et al. 1994; Gudnason et al. 1997; Inazu et al. 1994) and associations similar to those observed with the TaqIB polymorphism have been observed. The TaqIA polymorphism in intron 10 is in strong linkage disequilibrium with the TaqIB polymorphism in intron 1 (Kondo et al. 1989), and there is also significant linkage disequilibrium between several other polymorphisms (Kuivenhoven et al. 1997). No data on these other polymorphisms appear to contradict the findings with the TaqIB polymorphism.

The Variability Gene Concept

It may be plausible, and there are some supporting data in the literature (Groover et al. 1960), that the amount of variation in an individual's risk factors could be as important for his health prospects as absolute risk factor level (Berg 1994). We therefore developed a method to detect genes ("variability genes") that take part in determining the framework within which dietary or life style factors may cause risk factor variation. We made use of the fact that since monozygotia (MZ) twins share nuclear genes, any difference in risk factor level between the two members of a pair would be caused by dietary or life style factors. A permissive variability gene would be detected by larger within-pair difference in the variable under study in MZ pairs who possessed the gene than in those who lacked the gene. A restrictive variability gene would be detected by the opposite effect. In our initial studies (Berg 1983) we used random genetic markers and indeed found a variability gene effect of a blood group system that could easily be reproduced in a second series (Berg 1988). We concluded that the variability gene concept was probably valid. For practical purposes, we termed genes contributing to the framework within which dietary or life style factors can cause risk factor variation (without necessarily affecting absolute risk factor level), "variability genes," to distinguish them from "level genes" that only exhibit association with absolute risk factor level.

We proceeded to look for variability gene effect in polymorphisms at loci related to risk factors for CHD and found evidence for a variability effect of genes in apoB and apoA-I polymorphisms on lipids as well as body mass index (for review, see Berg 1994). In the years since we launched the variability gene concept (Berg 1986), variability gene effects have been reported by several other workers (Tikkanen et al. 1990; Reilly et al. 1991; Lopez-Miranda et al. 1994; Friedlander et al. 1995, 1997; Lee et al. 1995; Peacock et al. 1995; Humphries et al. 1996, 1997).

Although the variability gene phenomenon has been studied much less than the classical association between absolute risk factor level and specific genes, the existence of variability gene effects has been well confirmed.

The potential importance of the variability gene concept is that a person's total genetic risk for a disease such as CHD could be the result of his *combination* of level genes and variability genes. For example, a person could have variability genes permitting a high amount of variation in a risk factor such as cholesterol and at the same time have level genes that would tend to increase the absolute risk factor level. Theoretically, this could be a situation where a pro-atherogenic profile easily develops if dietary and life style precautions are not taken. The situation would be opposite for a person with restrictive variability genes and level genes that tend to reduce absolute risk factor level. Some persons in this situation could presumably maintain an anti-atherogenic lipid profile even with substantial intake of fats and calories.

"Variability Gene" Effect of Normal CETP Genes

We examined the TaqIB polymorphism in the CETP gene also with respect to the variability gene effect and found significantly higher within-pair differences in total cholesterol and LDL cholesterol (LDLC) in MZ pairs having the allele than in pairs lacking the B1 allele (presence of restriction site) in the TaqIB polymorphism (Table 2; Berg et al. 1989). The difference between genotypes was impressive, with the within-pair difference in B1B1 homozygotes being approximately 2.5 times larger than in B2B2 homozygotes for LDLC. B1B1 homozygotes had more than twice the within-pair difference of B2B2 homozygotes in total cholesterol. This polymorphism had no significant effect on absolute level of either total cholesterol or LDLC. No other variability gene effect of this polymorphism

Table 2. Sex and age adjusted within-pair difference (Δ) in total and LDL cholesterol, in 145 pairs of monozygotic (MZ) twins of different genotypes in the TaqIB polymorphism at the CETP locus. (Extracted from Berg et al. 1989)

Genotype	No. of pairs	Δ total cholesterol (mmol/l)	Δ LDL cholesterol (mmol/l)
B1B1	45	1.07	0.87[a]
B1B2	76	0.70[b]	0.67[c]
B2B2	24	0.46[d]	0.36

[a] Significance of difference from B2B2 group: p = 0.007
[b] Significance of difference from B1B1 group: p = 0.01
[c] Significance of difference from B2B2 group: p = 0.02
[d] Significance of difference from B1B1 group: p = 0.002

was observed. In particular, there was no variability gene effect on HDLC or apoA-I. Thus, we were faced with a situation where a polymorphism had a level gene effect but not a variability gene effect on HDLC and apoA-I, as well as a variability gene effect on total cholesterol and LDLC, but not on HDLC or apoA-I. Both the level gene and the variability gene effects of alleles in this polymorphism may be clinically relevant.

In a study on the lipoprotein response to a lipid-lowering diet in patients with type 1 diabetes, Dullaart et al. (1997) found that VLDL and LDL fell only in B1B1 homozygotes, apparently confirming the permissive variability effect of the B1 allele that we had detected.

A Case/Control Study of Myocardial Infarction and the TaqIB Polymorphism at the CETP Locus

We have examined the TaqIB polymorphism at the CETP locus in a series of 220 Norwegian patients who had suffered myocardial infarction prior to age 60 and in 146 healthy controls. The distribution of genotypes in the CETP TaqIB polymorphism did not differ significantly between patients and controls (Table 3). Although the series were reasonably large, conclusions must be drawn with caution. Several of the patients had had their first heart attack several years before they were recruited to the study. The genotypes of patients who died are unknown.

The distribution of TaqIB genotypes in 36 patients who died after entering the study did not differ significantly from the distribution of genotypes in survivors. This finding seems to suggest that people with genotype B2B2, despite their favorable HDL level, have no protection against death once they have suffered a heart attack. This may be at variance with findings in a study by Kuivenhoven et al. (1998), where the death rate in B2B2 persons treated with Pravastatin was zero. This finding differed from the death rate in B1B1 or B1B2 people who, however, also had low mortality over the two-year period of study (Kuivenhoven et al. 1998).

In our study, six of the seven B2B2 homozygotes who died did have additional factors that may have contributed to their death (Table 4). Three of the

Table 3. Distribution of healthy people and coronary heart disease (CHD) patients in a Norwegian case/control study, according to genotype in the TaqIB at the CETP locus (expected numbers calculated by the proportion method in brackets)

Group	No. with genotype			Total
	B1B1	B1B2	B2B2	
Healthy people	46	76	24	146
	(45.5)	(72.2)	(28.3)	
CHD patients	68	105	47	220
	(68.5)	(108.8)	(42.7)	
Both groups	114	181	71	366

$\varkappa^2 = 1.4$, 2 D.F., N.S.

Table 4. Lp(a) lipoprotein (Lp(a)) level, apoE genotype and smoking status in B2B2 homozygotes in the TaqIB polymorphism at the CETP locus who have died after entering a case/control study of myocardial infarction (MI) in Norway

Identification number	Gender	Lp(a) level (mg/dl)	ApoE genotype	Ever smoker	Continued to smoke after MI
14064	Male	10.6	3–4	Yes	No
14077	Male	0	3–3	?	
14096	Male	3.5	3–3	Yes	Yes
14121	Male	31.6[a]	3–4	Yes	Yes
14143	Female	28.6[a]	3–4	No	
14223	Male	27.3[a]	3–4	Yes	No
14240	Male	15.0[b]	3–3	Yes	Yes

[a] Above 90th centile in the healthy population
[b] Above 80th centile in the healthy population

seven persons had continued to smoke after the heart attack, four had high or very high Lp(a) lipoprotein levels (Table 4) and four had the apoE4 variant. Five of the seven B2B2 homozygotes who had died had high Lp(a) lipoprotein levels and/or had continued to smoke. Thus, it appears that B2B2 homozygotes who died did have serious risk factors that might have outweighed any protective effect of the B2B2 genotype. I conclude that this study does not exclude a favorable effect of the B2 gene, with respect to CHD.

Normal Polymorphism at the CETP Locus and Disease Risk

Several workers have analyzed disease risk in relationship to normal polymorphism at the CETP locus, and evidence has emerged that there are differences with respect to CID risk between genotypic groups.

Zhong et al. (1996) found increased prevalence of CHD in carriers of the D442G mutation in exon 15 of the CETP gene compared to Japanese-American men not carrying the mutation. This mutation is present in 5.2 % of Japanese-American men and, therefore, by definition forms part of a normal genetic polymorphism. The odds ratio (OR) for CHD in mutation carriers, after correction for known risk factors and HDL level, was 1.61.

Bruce et al. (1998), studying the I405V polymorphism at the CETP locus in Japanese men, found no difference in CHD prevalence between genotypes in the total series. However, in a sub-population with hypertriglyceridemia, the prevalence differed significantly between genotypes (38 %, 27 % and 18 %, respectively; $p < 0.05$). Thus, association between normal polymorphism at the CETP locus and CHD may be important in sub-groups with different metabolic status.

Based on a study of the I405V polymorphism at the CETP locus, Gudnason et al. (1997) argued that non-smoking men who reported alcohol consumption and who were homozygous for the 405V allele could have from 32 % to 40 % lower risk of having a heart attack than people carrying the I405 allele.

Tenkanen et al. (1991), studying three polymorphisms, did not observe significant differences in CETP allele distribution between healthy people, hyperlipidemic subjects and persons with myocardial infarction. However, the last group comprised only 72 individuals and conclusions must be drawn with caution.

A recent paper by Kuivenhoven et al. (1998) has been the focus of much attention. These workers studied the effect of Pravastatin on progression of coronary atherosclerosis and found that the pathological process progressed significantly more slowly in persons with genotype B1B1 or B1B2 than in B2B2 homozygotes (Table 5). Their study indicated that Pravastatin had no beneficial effect on the atherogenic process in the 16 % of patients who were B2B2 homozygotes. The favorable effect observed in B1B1 homozygotes and B1B2 heterozygotes could be reminiscent of the permissive variability gene effect of the B1 allele (Berg et al. 1989) on cholesterol. However, the difference between genotypic groups in the rate of progression of the atherogenic process did not seem to be mediated through an effect of Pravastatin on lipids (Table 5; Kuivenhoven et al. 1998).

The study by Kuivenhoven et al. (1998) would seem to suggest that people with the B2B2 genotype were particularly at risk for atherosclerosis and that there was little one could do once CHD had become manifest. The authors recorded two morphometric variables: "mean luminal diameter" and "minimal luminal diameter." The measurements (in mm) clearly showed that both parameters diminished during the two years of study and significantly more in B2B2 individuals than in B1B1 persons, in the Pravastatin group (Table 5).

However, close scrutiny of the data shows that the final size of the lumen (after correcting base line measurements with the changes over the two-year period presented in the paper) differed modestly between genotypes in the Pravastatin treatment group. The "mean luminal diameter" in people with genotype B2B2 was merely 0.03 mm narrower than that in B1B1 homozygotes and only 0.02 mm narrower than in B1B2 heterozygotes; corresponding figures for "minimal luminal diameter" are 0.07 and 0.06 mm. Thus, despite the significant differ-

Table 5. Effect of Pravastatin treatment on progression of coronary atherosclerosis, according to genotype in the TaqIB polymorphism at the CETP locus. (Extracted from Kuivenhoven et al. 1998)

Variable	Mean in people with genotype		
	B1B1	B1B2	B2B2
Decrease in mean luminal diameter (mm)	0.05	0.07	0.09
Decrease in minimal luminal diameter (mm)	0.01	0.04	0.04
Base line HDL cholesterol (mg/dl)	34	36	39
Increase in HDL cholesterol (mg/dl)	4.3	3.9	3.5
Base line LDL cholesterol (mg/dl)	167	166	169
Decrease in LDL cholesterol (mg/dl)	56.8	49.5	56.5
Myocardial infarction or death from cardiovascular causes (%)	2.9	1.9	0

ence between genotypes with respect to progression of the atherogenic process over a two-year period, the difference between B2B2 homozygotes and people possessing the B1 gene was unimpressive in the Pravastatin group. On the other hand, the findings in the Pravastatin group did not suggest that B2B2 homozygotes had less risk than people with other genotypes; all genotypic groups had a similar degree of arterial narrowing by the end of the study.

Interestingly, mortality was zero in B2B2 homozygotes in the Pravastatin-treated group (it was also low in people with B1B1 or B1B2 genotypes).

Normal Genes at the CETP Locus and Protection Against Atherosclerosis

Several examples of association between CHD and normal genes at the CETP locus are discussed above. If genes in polymorphisms are associated with disease, their alleles could be protective against the disease. Is there evidence that normal genes at the CETP locus are protective with respect to CHD?

Possessing a gene that can be identified as the B2 allele in the TaqIB polymorphism should theoretically be advantageous because it is associated with higher HDLC and apoA-I levels in people who do not smoke (Kondo et al. 1989). However, the B2 gene is restrictive with respect to total cholesterol or LDLC variability (Berg et al. 1989). This would be a disadvantage in people with level genes that cause higher absolute levels of these variables, but would be advantageous if level genes cause lower total cholesterol and LDLC levels. Thus, the risk situation is complicated, depending to an extent also on genes at other loci, smoking status, alcohol consumption and probably metabolic context (Fielding and Havel 1996; Tall 1993).

Ukkola et al. (1994) observed a prevalence of 6 % in B2B2 homozygotes and of 21 % in B1B1 homozygotes of cerebrovascular disease in patients with non-insulin-dependent diabetes mellitus. The difference was statistically significant ($p < 0.02$) and indicates that genotype B2B2 may confer a degree of protection against cerebrovascular disease.

In a four-center study, Fumeron et al. (1995) uncovered interaction between the TaqIB polymorphism at the CETP locus and alcohol consumption with an influence on risk of myocardial infarction. In B2B2 homozygotes who consumed 50–75 g per day of alcohol, they observed a significant reduction in odds ratio (OR = 0.56) for myocardial infarction compared to people possessing the B1 gene. The OR in B2B2 persons reporting consumption of more than 75 g alcohol per day was 0.34. No such effect was seen in non-drinkers or in persons consuming less than 25 g per day. The two-thirds reduction in OR for myocardial infarction in B2B2 homozygotes who consumed more than 75 g per day of alcohol is strong support for the hypothesis of a protective effect of the B2B2 genotype against arteriosclerotic diseases.

In the study by Kuivenhoven et al. (1998), the most striking observations were made in the placebo group. Table 6 shows mean luminal diameter after two years in the placebo group (arrived at by adjusting the base line measurements for the changes during the two-year period reported by Kuivenhoven et al. 1998). Table 7 shows corresponding data for minimal luminal measurements. People with genotype B2B2 had the highest measurements (the widest vessel lumen) at the end of the study of "mean luminal diameter" as well as "minimal luminal diameter" (Tables 6 and 7). Thus, presence of the B2 allele was associated with slower progression of the atherogenic process. The findings in the placebo group of Kuivenhoven et al. (1998) confirm that a normal CETP gene, detected as the B2 allele in the TaqIB polymorphism confers a level of protection against atherosclerotic disease.

The accelerated atherogenesis in B2B2 homozygotes in the Pravastatin group as opposed to the placebo group suggests the possibility that the drug made the disease process worse in the B2B2 homozygotes. Statin treatment is known to

Table 6. Mean luminal diameter (mm) in total series (n = 807) at base line and after two years in the placebo group (n = 396) according to genotype in the TaqIB polymorphism in the CETP gene. (Extracted from Kuivenhoven et al. 1998)

Measurement	Mean luminal diameter in men with genotype		
	B1B1	B1B2	B2B2
Before treatment	2.80	2.81	2.81
After two years in placebo group	2.66	2.71	2.76

Table 7. Minimal luminal diameter (mm) in total series (n = 807) at base line and after two years in placebo group (n = 396) according to genotype in the TaqIB polymorphism in the CETP gene. (Extracted from Kuivenhoven et al. 1998)

Measurement	Median minimal luminal diameter in men with genotype		
	B1B1	B1B2	B2B2
Before treatment	1.90	1.92	1.86
After two years in placebo group	1.77	1.83	1.85

reduce transfer of esterified cholesterol from HDL (Ahnadi et al. 1993; Guerin et al. 1995). It is not implausible that a change in reverse cholesterol transport with accumulation of lipids in peripheral cells could unfavourably affect the disease process in the arterial wall in the metabolic conditions of B2B2 homozygotes.

Kuivenhoven et al. (1998) suggested that the 16% of patients who are B2B2 homozygotes may not be given statins because they do not benefit from it with respect to the pathological process itself. The comparison between their placebo group and Pravastatin group suggests that Pravastatin may even accelerate the atherogenic process in B2B2 persons.

The TaqIB polymorphism is by far the most thoroughly studied polymorphism at the CETP locus. Studies of this polymorphism strongly suggest that there is indeed a protective effect of a gene (or genes) detectable as the B2 allele.

Conclusion

Even though the interrelationship between normal polymorphism at the CETP locus, HDL level, CETP level and atherogenesis may be complicated, I conclude that a normal gene (or genes) at the CETP locus detectable as the B2 allele in the TaqIB polymorphism confers protection to its bearers. This protection may, however, be modified by life style factors such as smoking or alcohol, or by genes at other loci. The level gene or the variability gene effect of alleles at this locus should be taken into account in attempts to fully understand how normal genes at the CETP locus may protect against atherosclerotic disease.

Acknowledgments

Research in the author's laboratory is supported by grants from The Norwegian Council on Cardiovascular Diseases and Anders Jahre's Foundation for the Promotion of Science.

References

Ahnadi C-E, Berthezene F, Ponsin G (1993) Simvastatin-induced decrease in the transfer of cholesterol esters from high density lipoproteins to very low and low density lipoproteins in normolipidemic subjects. Atherosclerosis 99:219–228

Arai T, Yamashita S, Sakai N, Hirano K, Okada S, Ishigami M, Maruyama T, Yamane M, Kobayashi H, Nozaki S, Funahashi T, Kameda-Takemura K, Nakajima N, Matsuzawa Y (1996) A novel nonsense mutation (G181X) in the human cholesteryl ester transfer protein gene in Japanese hyperalphalipo-proteinemic subjects. J Lipid Res 37:2145–2154

Berg K (1983) Genetics of coronary heart disease. In: Steinberg AG, Bearn AG, Motulsky AG, Childs B (eds) Progress in medical genetics, New Series V. Saunders, Philadelphia, pp. 35–90

Berg K (1986) Normal genetic lipoprotein variations and atherosclerosis. In: Sirtori CR, Nichols AV, Franceschini G (eds) Human apolipoprotein mutants. Impact on atherosclerosis and longevity. Plenum Press, New York and London, pp. 31–49

Berg K (1988) Variability gene effect on cholesterol at the Kidd blood group locus. Clin Genet 33:102–107

Berg K (1993) Genetic and environmental factors in the development of cardiovascular disease. In: Galteau M-M, Siest G, Henry J (eds) Biologie prospective. Comptes rendus du 8^e Colloque de Pont-a-Mousson. John Libbey Eurotext, Paris, pp. 471–480

Berg K (1994) Gene-environment interaction: variability gene concept. In: Goldbourt U, de Faire U, Berg K (eds) Genetic factors in coronary heart disease. Kluwer Academic Publishers, Dordrecht/Boston/London, pp. 373–383

Berg K, Kondo I, Drayna D, Lawn R (1989) "Variability gene" effect of cholesteryl ester transfer protein (CETP) genes. Clin Genet 35:437–445

Bernard S, Moulin P, Lagrost L, Picard S, Elchebly M, Ponsin G, Chapuis F, Berthezene F (1998) Association between plasma HDL-cholesterol concentration and Taq1B CETP gene polymorphism in non-insulin-dependent diabetes mellitus. J Lipid Res 39:59–65

Bruce C, Sharp DS, Tall AR (1998) Relationship of HDL and coronary heart disease to a common amino acid polymorphism in the cholesteryl ester transfer protein in men with and without hypertriglyceridemia. J Lipid Res 39:1071–1078

Bu X, Warden CH, Xia YR, De Meester C, Puppione DL, Teruya S, Lokensgard B, Daneshmand S, Brown J, Gray RJ, Rotter JI, Lusis AJ (1994) Linkage analysis of the genetic determinants of high density lipoprotein concentrations and composition: evidence for involvement of the apolipoprotein A-II and cholesteryl ester transfer protein loci. Hum Genet 93:639–648

Drayna D, Lawn R (1987) Multiple RFLPs at the human cholesteryl ester transfer protein (CEPT) locus. Nucleic Acids Res 15:4698

Drayna D, Jarnagin As, McLean J, Henzel W, Kohr W, Fielding C, Lawn R (1987) Cloning and sequencing of human cholesteryl ester transfer protein cDNA. Nature 327:632–634

Dullaart RP, Hoogenberg K, Riemens SC, Groener JE, van Tol A, Sluiter WJ, Stulp BK (1997) Cholesteryl ester transfer protein gene polymorphism is a determinant of HDL cholesterol and of the lipoprotein response to a lipid-lowering diet in type 1 diabetes. Diabetes 46:2082–2087

Dullaart RP, Deusekamp BJ, Riemens SC, Hoogenberg K, Stulp BK, van Tol A, Sluiter WJ (1998) High-density lipoprotein cholesterol is related to the TaqIB cholesteryl ester transfer protein gene polymorphism and smoking, but not to moderate alcohol consumption in insulin-dependent diabetic men. Scand J Clin Lab Invest 58:251–258

Fielding CJ, Havel RJ (1996) Cholesteryl ester transfer protein: friend or foe? J Clin Invest 97:2687–2688

Freeman DJ, Packard CJ, Shepherd J, Gaffney D (1990) Polymorphisms in the gene coding for cholesteryl ester transfer protein are related to plasma high-density lipoprotein cholesterol and transfer protein activity. Clin Sci 79:575–581

Freeman DJ, Griffin BA, Holmes AP, Lindsay GM, Gaffney D, Packard CJ, Shepherd J (1994) Regulation of plasma HDL cholesterol and subfraction distribution by genetic and environmental factors. Associations between the TaqIB RFLP in the CETP gene and smoking and obesity. Arterioscler Thromb 14:336–344

Friedlander Y, Berry EM, Eisenberg S, Stein Y, Leitersdorf E (1995) Plasma lipids and lipoproteins response to a dietary challenge: analysis of four candidate genes. Clin Genet 47:1–12

Friedlander Y, Austin MA, Newman B, Edwards K, Mayer-Davis EJ, King M-C (1997) Heritability of longitudinal changes in coronary-heart-disease risk factors in women twins. Am J Human Genet 60:1502–1512

Fumeron F, Betoulle D, Luc G, Behague I, Ricard S, Poirier O, Jemaa R, Evans A, Arveiler D, Marques-Vidal P, Bard J-M, Fruchart J-C, Ducimetiere P, Apfelbaun M, Cambien F (1995) Alcohol intake modulates the effect of a polymorphism of the cholesteryl ester transfer protein gene on plasma high density lipoprotein and the risk of myocardial infarction. J Clin Invest 96:1664–1671

Groover ME, Jernigan JA, Martin CD (1960) Variations in serum lipid concentration and clinical coronary disease. Am J Med Sci 53:133–139

Gudnason V, Thormar K, Humphries SE (1997) Interaction of the cholesteryl ester transfer protein I405V polymorphism with alcohol consumption in smoking and non-smoking healthy men, and the effect on plasma HDL cholesterol and apoAI concentration. Clin Genet 51:15–21

Guerin M, Dolphin PJ, Talussot C, Gardette J, Berthezene F, Chapman MJ (1995) Pravastatin modulated cholesteryl ester transfer from HDL to apoB-containing lipoproteins and lipoprotein subspecies profile in familial hypercholesterolemia. Arterioscler Thromb Vasc Biol 15:1359–1368

Hannuksela ML, Liinamaa MJ, Kesaniemi YA, Savolainen MJ (1994) Relation of polymorphisms in the cholesteryl ester transfer protein gene to transfer protein activity and plasma lipoprotein levels in alcohol drinkers. Atherosclerosis 110:35–44

Humphries SE, Talmud PH, Cox C, Sutherland W, Mann J (1996) Genetic factors affecting the consistency and magnitude of changes in plasma cholesterol in response to dietary challenge. Quart J Med 89:671–680

Humphries SE, Thomas A, Montgomery HE, Green F, Winder A, Miller G (1997) Gene-environment interaction in the determination of plasma levels of fibrinogen. Fibrinolysis Proteolysis 11: (suppl. 1) 3–7

Inazu A, Jiang XC, Haraki T, Yagi K, Kamon N, Koizumi J, Mabuchi H, Takeda R, Takata K, Moriyama Y (1994) Genetic cholesteryl ester transfer protein deficiency caused by two prevalent mutations as a major determinant of increased levels of high density lipoprotein cholesterol. J Clin Invest 94:1872–1882

Kauma H, Savolainen MJ, Heikkila R, Rantala AO, Linja M, Reunanen A, Kesaniemi YA (1996) Sex difference in the regulation of plasma high density lipoprotein cholesterol by genetic and environmental factors. Human Genet 97:156–162

Kondo I, Berg K, Drayna D, Lawn R (1989) DNA polymorphism at the locus for human cholesteryl ester transfer protein (CETP) is associated with high density lipoprotein cholesterol and apolipoprotein levels. Clin Genet 35:49–56

Kuivenhoven JA, de Knijff P, Boer JM, Smalheer HA, Botma GJ, Seidell JC, Kastelein JJ, Pritchard PH (1997) Heterogeneity at the CETP gene locus. Influence on plasma CETP concentrations and HDL cholesterol levels. Arterioscler Thromb Vasc Biol 17:560–568

Kuivenhoven JA, Jukema JW, Zwinderman AH, de Knijf P, McPherson R, Bruschke AVG, Lie KI, Kestelein JJP (1998) The role of a common variant of the cholesteryl ester transfer protein gene in the progression of coronary atherosclerosis. N Engl J Med 338:86–93

Lee E, Tu L, Abdolell M, Corey P, Connelly PW, Jenkins DJA, Hegele RA (1995) Evidence for gene-diet interaction in the response of plasma lipoproteins to dietary fibre. Nutr Metab Cardiovasc Dis 5:261–268

Lopez-Miranda J, Ordovas JM, Espino A, Marin C, Salas J, Lopez-Segura F, Jimenez-Pereperez J, Perez-Jimenez F (1994) Influence of mutation in human apolipoprotein A-1 gene promoter on plasma LDL cholesterol response to dietary fat. Lancet 343:1246–1249

Lusis AJ, Zollman S, Sparkes RS, Klisak I, Mohandas T, Drayna D, Lawn RM (1987) Assignment of the human gene for cholesteryl ester transfer protein to chromosome 16q12–16q21. Genomics 1:232–235

Marotti KR, Castle CK, Boyle TP, Lin AH, Murray RW, Melchior GW (1993) Severe atherosclerosis in transgenic mice expressing simian cholesteryl ester transfer protein. Nature 364:73–75

Mendis S, Shepherd J, Packard CJ, Gaffney D (1990) Genetic variation in the cholesteryl ester transfer protein and apolipoprotein A-I genes and its relation to coronary heart disease in a Sri Lankan population. Atherosclerosis 83:21–27

Mitchell RJ, Earl L, Williams J, Bisucci T, Gasiamis H (1994) Polymorphisms of the gene coding for the cholesteryl ester transfer protein and plasma lipid levels in Italian and Greek migrants to Australia. Human Biol 66:13–25

Nora JJ, Lortscher RH, Spangler RD, Nora AH, Kimberling WJ (1980) Genetic-epidemiologic study of early-onset ischemic heart disease. Circulation 61:503–508

Nora JJ, Berg K, Nora AH (1991) Cardiovascular diseases. Genetics, epidemiology and prevention. Oxford University Press, New York.

Peacock RE, Karpe F, Talmud PJ, Hamsten A, Humphries SE (1995) Common variation in the gene for apolipoprotein B modulates postprandial lipoprotein metabolism: a hypothesis generating study. Atherosclerosis 116:135–145

Reilly SL, Ferrell RE, Kottke BA, Kamboh MI, Sing CF (1991) The gender-specific apolipoprotein E genotype influence on the distribution of lipids and apolipoproteins in the population of Rochester, MN. I. Pleiotropic effects on means and variances. Am J Human Genet 49:1155–1166

Song GJ, Han GH, Namkoong Y, Lee HK, Park YB, Lee CC (1997) The effects of the cholesteryl ester transfer protein gene and environmental factors on the plasma high density lipoprotein cholesterol level in the Korean population. Mol Cells 7:615–619

Tall A (1993) Plasma lipid transfer proteins. Annu Rev Biochem 64:235–257

Tenkanen H, Koshinen P, Kontula K, Aalto-Setala K, Manttari M, Manninen V, Runeberg SL, Taskinen NR, Ehnholm C (1991) Polymorphisms of the gene encoding cholesterol ester transfer protein and serum lipoprotein levels in subjects with and without coronary heart disease. Human Genet 87:574–578

Tikkanen MJ, Xu C-F, Hämäläinen T, Talmud P, Sarna S, Huttunen JK, Pietinen P, Humphries S (1990) XbaI polymorphism of the apolipoprotein B gene influences plasma lipid response to diet intervention. Clin Genet 37:327–334

Ukkola O, Savolainen MJ, Salmela PI, von Dickhoff K, Kesäniemi YA (1994) DNA polymorphisms at the locus for human cholesteryl ester transfer protein (CETP) are associated with macro- and microangiopathy in non-insulin-dependent diabetes mellitus. Clin Genet 46:217–227

Zhong S, Sharp DS, Grove JS, Bruce C, Yano K, Curb JD, Tall AR (1996) Increased coronary heart disease in Japanese-American men with mutation in the cholesteryl ester transfer protein gene despite increased HDL levels. J Clin Invest 97:2917–2923

Does the Gene Encoding Apolipoprotein A–I$_{Milano}$ Protect the Heart?

C. R. Sirtori, L. Calabresi

Introduction

Cardiac protection is generally believed to be exerted by high density lipoproteins (HDL). These lipoproteins can, in fact, remove cholesterol from the arterial wall, thus preventing plaque formation. Plasma levels of HDL are elevated in conditions where the atherosclerosis risk is reduced: women in fertile age, physically fit people, nonsmokers, etc. Conversely, HDL levels are reduced in smokers, diabetics and generally in all conditions leading to an enhanced risk of atherosclerotic vascular disease (Gordon and Rifkind 1989).

The discovery of the first mutation of the human proteins responsible for cholesterol transport, i.e., apolipoprotein A-I (apoA-I), the major component of HDL, was, on the one hand, a major step forward in biochemistry and genetics, and on the other, carried a conflicting message in terms of cardiovascular prevention. This mutant, the A-I$_{Milano}$ apolipoprotein, shows, in fact, potentially improved mechanistic properties, but also points out the ambiguous nature of human genetics, leading to lower levels of a putatively protective circulating protein in the face of reduced cardiovascular risk.

A-I$_{Milano}$ – The Clinical Observation

A 42-year-old man with a clinical history of gastritis was referred to us for consultation in 1975, because of an apparently complex lipoprotein disorder (Franceschini et al. 1980). This disorder was characterized by:

1) hypertriglyceridemia, ranging between 300 and 1,000 mg/dl (3.40 to 11.30 mM/L); mildly to markedly elevated cholesterolemia (range 240–412 mg/dl) (6–10.5 mM/L);
2) apparent resistance to the, at the time, major lipid-lowering medication, clofibrate. His doctor's reports indicated that the subject responded with a rise instead of a fall in triglyceridemia.

Due to a prior accident at his work (railroad employee), he had a mild form of depression, with occasional flareups. Questions about family history also provided little information of interest. His mother, aged 70, had suffered a stroke and was disabled. No other family member had a history of atherosclerotic vascular disease.

V. Boulyjenkov, K. Berg, Y. Christen (Eds.)
Genes and Resistance to Diseases
© Springer-Verlag Berlin Heidelberg 2000

Evaluation of HDL cholesterol levels (not routine at that time) by selective precipitation and later by ultracentrifugation showed ranges between 7 and 13 mg/dl (0.2–0.4 mM/L).

In view of the apparent discrepancy between the biochemical and the clinical data, the subject underwent an extensive evaluation which also focussed on other potential determinants of the lipoprotein disorder, i.e., liver changes, a cancer-related condition, and others. Interestingly, two of the three children of the man, aged between 12 and 16 years, proved to have an essentially identical lipoprotein profile, as did his father who at the time was around 75 years of age. However, his mother, who clearly suffered from an atherosclerotic condition, had normal lipid/lipoprotein levels (Franceschini et al. 1981). Frustration with the difficulty of finding an acceptable diagnosis was compounded by the clear lack of interest of numerous scientists who were contacted for advice. Mutations in apolipoproteins were, as yet, believed not to exist, in view of the very narrow structural requirements for lipoprotein formation (Tall et al. 1975).

Finally, after four years of difficulties, a successful conclusion was reached. It could in fact be shown that apoA-I in the proband was characterized by the presence of a cysteine residue. Circulating dimer forms ($A-I_{Milano}/A-I_{Milano}$) or complexes with apoA-II, also containing a cysteine ($A-I_{Milano}/A-II$), were evident in SDS electrophoretic separations in the absence of a reducing agent, such as mercaptoethanol (Weisgraber et al. 1980). This finding proved that the subject under investigation was the carrier of the first mutation identified in circulating apolipoproteins up to that moment.

The major traits of the mutant, from then on named apoA-I_{Milano}, were:
1) low HDL cholesterol (range, in the five carriers, between 7 and 18 mg/dl);
2) variable hypertriglyceridemia (three of five showed hypertriglyceridemia);
3) normal to moderately elevated total and low density lipoprotein (LDL) cholesterolemia;
4) non-sex-linked, probably autosomal disorder.

Shortly thereafter, the site of the mutation was established as a substitution of cysteine for arginine at position 173 in the apoA-I protein chain (Weisgraber et al. 1983).

The Limone sul Garda Link – The Genetics of ApoA-I$_{Milano}$

Crucial help in defining the unique biological properties of apoA-I_{Milano} ($A-I_M$) and direct information about its genetic transmission came from a survey of the population of Limone sul Garda, a small village on Lake Garda, where the $A-I_M$ proband was born. Limone provided a rather unique opportunity because the village (approximately 1,000 citizens) had been separated from neighboring villages by the lack of incoming roads until the last world war. For this reason, the Limone population had a high rate of inbreeding and the mutant gene became established through the last three centuries.

Detailed analysis of the genealogic tree, built using church records, provided clear evidence that the first expression of the gene occurred in a couple married

Table 1. A–I$_{Milano}$ carriers

Total: 37
Age: 3–93 (41 ± 23)
M = 23; F = 14
HTG = 11
Age: 3–93 (52 ± 22)
TG: 262 ± 39; HDL-C: 13.3 ± 2.5
NTG = 26
Age: 7–87 (36 ± 21)
TG: 131 ± 42; HDL-C: 18.7 ± 7.8

in 1780 (Gualandri et al. 1985). Thereafter, some 80 obligate carriers were identified and 40 are alive today, 37 of whom are fully characterized (Table 1). Interestingly, obligate carriers in the past centuries seemed to be relatively long-lived and the only two early deaths, due to acute vascular conditions, in this century occurred in a 53-year-old man, a heavy smoker with arterial hypertension and, more recently, a 54-year-old man, also a heavy smoker.

All carriers, easily identifiable by a combination of an isoelectric focusing (IEF) procedure (Franceschini et al. 1981; Gualandri et al. 1985), show reduced HDL-cholesterol levels and, in some, variable expression of hypertriglyceridemia (Table 1), leading to a clear-cut type IV phenotype. In a small number of carriers, significant hypercholesterolemia could also be detected. It should be noted that none of the carriers (portatori) is on any diet and many are smokers. Indeed the present diet in Limone, where the major occupation is tourism, can be considered as rather atherogenic, being rich in calories and saturated fat and with a generous intake of ethanol.

A careful genetic evaluation of the population leads to the conclusion that transmission is direct without any generational jump. The phenotype has the same frequency in both sexes (χ^2 1.336), and the transmission occurs according to the Mendelian mode of autosomic codominant inheritance. In view of the apparent low incidence of atherosclerotic disease in the carriers, a global calculation of death risk in Limone sul Garda in a 10-year period was carried out by applying a revision of the WHO-ICD codes, and calculating the expected deaths from Italian five-year age-specific rates (Table 2). It can be noted that, globally, the Limone sul Garda population has an apparently reduced risk of coronary heart disease, and this certainly influences the apparently lower risk in the "portatori".

Two clinical studies were carried out in Limone sul Garda at different times. In the first one (1981–82), a series of "portatori" (n = 33) and relatives (n = 43) was not very informative (Gualandri et al. 1985). There was an apparent excess incidence of venous disease in the "portatori" and, from the clinical reports, there also appeared to be a slight tendency to bleeding episodes not found in close relatives.

This initial, strictly clinical study, carried out directly by the author, was followed by a more recent, detailed cardiovascular investigation. This investigation

Table 2. Death risk in Limone sul Garda in the 10-year period 1972–1981. Revision of WHO-ICD codes and expected deaths from Italian five-yr age-specific rates

	Observed (O)	Expected (E)	O–E (% of E)	P of χ^2
CHD	7	14.32	–51.1	(0.10 > P > 0.05)
Strokes	12	12.51	–4.1	ns[a]
Others	16	13.99	+14.4	ns
Cancer	18	17.76	+1.3	ns
[a]Non-significant (ns)				

made use of more advanced technologies and examined 21 carriers versus 42 non-carriers from the same kindred, age- and sex-matched. These subjects underwent, in addition to the clinical examination, measurement of the carotid intima-media thickness (IMT; Pignoli et al. 1986), which was recently shown to be a very sensitive indicator of atherosclerosis risk (O'Leary et al. 1999). In addition, the flow dynamics of forearm artery was evaluated by an arterial distensibility/compliance test (Baldassarre et al. 1995), and finally all the selected subjects underwent major cardiac dynamic testing (stress test, echo and others). In this more recent study, the carotid IMT and arterial distensibility were also evaluated *vis a vis* patients from our clinical center who had primary reductions of HDL cholesterol levels, in the absence of mutations and in many cases associated with coronary disease. This study, currently undergoing final evaluation, has shown that the A-I_M carriers have perfectly normal IMT, and also a normal arterial distensibility and cardiac dynamic parameters. Interestingly, in the same tests, patients with primary reduction of HDL cholesterol levels show significantly larger IMTs and reduced arterial distensibility versus controls.

These and other findings provided convincing evidence that the mutation was to a large extent responsible for the cardiovascular protective effect, and prompted a number of studies aimed at clarifying the mechanism/s whereby a mutated apoA-I in reduced amounts could carry out the "protective" role of apoA-I with improved effectiveness.

Studies of the Biochemical Properties of A-I_M and Potential Therapeutic Implications

A number of studies have clearly established novel biochemical features of the mutant of apoA-I. An early study has shown that the behavior of isolated *monomeric A-I_M* is characterized by (Franceschini et al. 1985):
1) accelerated association with model phospholipids in solution;
2) easier desorption of lipids from lipoprotein complexes upon exposure to dissociating agents.

This study provided the first experimental evidence supporting a role for a discontinuity in the α-helical segment containing the mutation (Fig. 1) in explaining

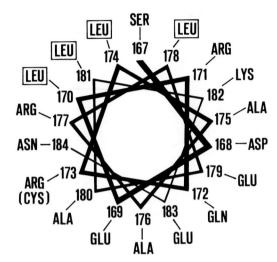

Fig. 1. Helical wheel presentation of the presumed α-helical structure of segment 167–184 of apoA-I. In apoA-I_M, the Arg[173] is substituted with cysteine. (Franceschini et al. 1985)

the physiological properties of the mutant. The discontinuity, brought about by the presence of the cysteine residue, will result in the loss of some of the hydrogen bonds keeping the α-helices together, thus leading to a looser protein structure, allowing faster binding to, and easier desorption, from lipids (Franceschini et al. 1985).

A metabolic study investigating the fate of the mutant in vivo was carried out at the National Institutes of Health in Washington. Both normal apoA-I and A-I_M isolated from plasma were labeled with either ^{125}I or ^{131}I and injected into recipients with a normal apolipoprotein profile, as well as into A-I_M carriers. Upon examination of the plasma radioactivity die away curves, both in whole plasma and within electrophoretically separated apolipoproteins, two conclusions were reached (Roma et al. 1993; Fig. 2):

1) apoA-I_M in the monomeric form is very rapidly cleared from plasma, thus probably accounting for the low steady state levels of the abnormal apolipoprotein, and also possibly accounting for its prompt activity on tissue lipid removal;

2) in contrast, the A-I_M dimer (A-I_M/A-I_M) is characterized, both in carriers and noncarriers, by an extremely delayed turnover, with a half-life approximately three-fold longer versus normal apoA-I and five-fold longer versus the A-I_M monomer.

Since in the "portatori" the majority of the mutant A-I is present in plasma in the dimeric form, it may be safely concluded that the rapidly turning-over monomer is probably responsible for the very efficient interaction with tissue lipids and, consequently, also for the reduced steady-state levels. In the case of the A-I_M dimer, possible mechanistic explanations are less clear. The in vitro stability of the HDL particles containing the A-I_M/A-I_M dimer has been well established (Franceschini et al. 1990). Furthermore, as reported below, the dimer itself shows

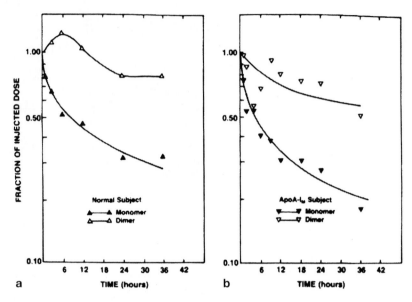

Fig. 2. Plasma radioactivity decay curves of apoA-I monomer and dimer in normal subject (A) and in A-I$_{Milano}$ subject (B). (Roma et al. 1993)

potent cholesterol-removing properties, in addition to other possibly beneficial protective mechanisms.

Apolipoprotein A-I/HDL: Biochemical Properties and Therapeutic Potential

HDL play a major role in reverse cholesterol transport (RCT), the process by which excess cholesterol in peripheral tissues, including the arterial wall, is esterified in plasma and transported to the liver for excretion. This function of HDL is believed to explain, at least in part, the strong inverse correlation between plasma HDL levels and coronary heart disease found in many prospective and case-control studies (Gordon and Rifkind 1989). ApoA-I seems to carry the major responsibility for the antiatherogenic activity of HDL, because of the multiple roles it plays in RCT. It is the preferential acceptor of cell cholesterol (Rothblat et al. 1992), acts as a cofactor for the lecithin:cholesterol acyltransferase (LCAT) enzyme (Jonas 1991), and acts as a ligand for a putative HDL receptor (Allan et al. 1993). In addition, apoA-I displays unique properties, not directly related to its major activities in RCT but possibly involved in HDL protection against vascular disease (Table 3). Mice and rabbits transgenic for apoA-I, and therefore with an increased number of circulating HDL particles, appear to be protected from the development of diet- or gene-induced atherosclerosis (Rubin et al. 1991; Paszty et al. 1994; Duverger et al. 1996). Even more strikingly, adenovirus-mediated transfer of the apoA-I gene into apo E-deficient mice remarkably inhibited neointima formation after endothelial denudation

Table 3. Anti-atherogenic and anti-thrombogenic properties of apoA-I. (From Sirtori et al. 1999)

Stimulation of reverse cholesterol transport
Inhibition of LDL oxidation
Protection against the cytotoxic effects of oxidized LDL
Prevention of cholesterol accumulation in macrophages
    by modified LDL
Prevention of cytokine-induced cell-cell interactions
Inhibition of neutrophil degranulation
Endotoxins neutralization
Prostacyclin stabilization
Stimulation of prostacyclin release
Activation of fibrinolysis
Inhibition of procoagulant activity
Modulation of complement function

(DeGeest et al. 1997), suggesting a direct protective effect of apoA-I on the arterial wall.

ApoA-I is synthesized both in the liver and in the intestine as pre-pro-apoA-I. The 18A-amino acid signal peptide is co-translationally cleaved and apoA-I appears in plasma as the 249 amino acid pro-protein, and is then processed to the mature form of 243 residues (Brewer et al. 1978). The most striking feature of the mature apoA-I sequence is the presence of internal repeat units of 11 or 22 amino acids, with the potential to assume an amphipathic helical structure (Segrest et al. 1992). When bound to lipids, as in discoidal synthetic HDL (sHDL), the amphipathic helices run from side to side of the disk (Wald et al. 1990), with charged residues facing the aqueous phase and hydrophobic residues facing the acyl chains of the phospholipid bilayer. Specific sequences within apoA-I, required for the interaction between apoA-I helices and LCAT or cell membrane, can induce cholesterol esterification and efflux.

Studies with sHDL, made with phospholipids and either A-I$_M$/A-I$_M$ or apoA-I, have clearly shown that A-I$_M$/A-I$_M$ has a limited conformational flexibility compared with the wild-type protein, resulting in a restricted HDL particle size heterogeneity and in a unique organization of the protein on the surface of the discoidal sHDL particles (Fig. 3; Calabresi et al. 1997a). The introduction of the interchain disulfide bridge in A-I$_M$/A-I$_M$ generates a novel structural domain, consisting of the C-terminal $\sim$40 residues, which loops out of the sHDL surface and may explain some of the unique functions of the disulfide-linked A-I$_M$ dimer (Calabresi et al. 1994, 1997a). sHDL particles containing A-/I$_M$/A-I$_M$ are more efficient than those with apoA-I in promoting cholesterol efflux from cells and are equally effective in inhibiting the cytokine-induced expression of adhesion molecules on endothelial cells (Calabresi et al. 1997b), whereas native HDL containing A-I$_M$/A-I$_M$ are more resistant to remodeling by the LCAT and CETP enzymes (Franceschini et al. 1990), and, as shown before, are cleared from the human circulation at a slower rate than apoA-I-containing particles (Roma et al. 1993). A-I$_M$/A-I$_M$ would thus behave as a stable form of apoA-I, which might explain its powerful antiatherogenic activity in animal models (Ameli et al. 1994; Soma et al. 1995; Shah et al. 1998) and, possibly, in humans.

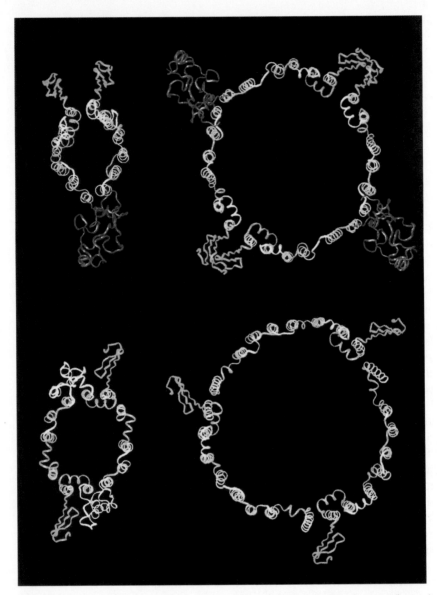

Fig. 3. Hypothetical model of synthetic HDL containing A-I$_M$/A-I$_M$ (top) and apoA-I (bottom). The discoidal sHDL are viewed from the top and are filled with phospholipid molecules (not shown). The ribbon representations of A-I$_M$/A-I$_M$ and apoA-I are colored as follows: lipid-binding core, light green; N-terminal domain, blue; A-I$_M$/A-I$_M$ C-terminal domain, red; apoA-I hinged domain, light yellow.

Development of Recombinant apoA-I and apoA-I$_M$ and Evaluation of Activity: Lessons from a Transgenic Mouse Model

Human apoA-I has been expressed in several eukaryotic cell lines. The propeptide is generally removed by an unidentified proteolytic process, resulting in the secretion of mature apoA-I, part of which is assembled with lipids into HDL-like particles (Sorci-Thomas et al. 1996; Forte et al. 1990) that need to be delipidated to obtain the lipid-free protein. Expression levels in eukaryotic cells are generally low, thus making it difficult to purify substantial amounts of apoA-I.

Most of these problems have been solved by keeping the pro-segment of the apoA-I sequence, thus producing pro-apoA-I instead of the mature protein (Moguilevski et al. 1989, 1993; Pyle et al. 1996; Holvoet et al. 1996). In this way, expression levels up to 6 mg/l were achieved (McGuire et al. 1996). The recombinant pro-apoA-I, which is efficiently converted into mature apoA-I in vivo (Saku et al. 1993), proved effective in raising plasma HDL levels and promoting RCT in humans (Carlson 1995). An alternative approach has been that of overexpressing apoA-I in E. coli by introducing silent mutations in the codons for the first apoA-I residues, or in adding the sequence coding for a bacterial signal peptide immediately upstream of the mature apoA-I sequence (Moguilevsky et al. 1993). This latter procedure was successfully applied to the industrial production of A-I$_M$/A-I$_M$, achieving expression levels up to 2 g/l in bioreactors of 75 liters (Calabresi et al. 1994).

A major issue in the development of apolipoprotein pharmaceuticals is the formulation of the recombinant proteins. Human apoA-I circulates in plasma mainly as a component of HDL; only a minor fraction ($<5\%$) is present as free protein, being released from HDL during metabolic remodeling (Liang et al. 1995). Locally produced lipid-free apoA-I can recruit phospholipids and cholesterol from cell membranes and may contribute to RCT (Oram and Yokohama 1996). However, metabolically derived or exogenous, i.v.-infused lipid-free apoA-I is removed from the circulation at a much faster rate than the HDL-bound protein (Nanjee et al. 1996). Therefore, apoA-I should be administered as a component of an HDL-like particle to achieve effective plasma levels.

Three studies have examined the effects of i.v.-injected, wild-type apoA-I against diet-induced atherosclerosis in animals. Results in all three studies were apparently positive for an antiatherosclerotic activity, but variability of results does not allow definitive conclusions to be reached (Trachtenberg et al. 1993; Mezdour et al. 1995; Miyazaki et al. 1995).

Four animal studies have been conducted with sHDL containing the recombinant A-I$_M$/A-I$_M$ (A-I$_M$/A-I$_M$-sHDL). The i.v. administration of A-I$_M$/A-I$_M$sHDL (40 mg of protein), given five times according to a schedule of evenly spaced injections before and after angioplasty, reduced by more than 70% the restenosis in cholesterol-fed rabbits (Ameli et al. 1994). The same sHDL were very effective in preventing intimal thickening induced in rabbit carotid by periarterial manipulation (Soma et al. 1995); this effect occurred without any changes in plasma cholesterol levels and was associated with a marked reduction of smooth muscle cell proliferation (Soma et al. 1995). In a more recent study, A-I$_M$/A-I$_M$sHDL injections (40 mg/kg of protein every other day for five weeks) prevented the progression of aortic atherosclerosis and reduced lipid and macrophage content of plaques in apoE-deficient mice, again despite no effects on plasma cholesterol (Shah et al. 1998). Altogether these data indicate that injections of sHDL containing the recombinant A-I$_M$/A-I$_M$ can bring about significant changes in arterial cell dynamics, leading to both plaque stabilization and lesion regression. Finally, very recently a study in rats provided a clear indication that apoA-I$_M$ dimer given i.v. at a dose of 20 mg/kg daily for 4–10 days after a thrombogenic insult can markedly delay thrombus formation, inhibit platelet aggregation and reduce thrombus weight (Li et al. 1999).

A transgenic model of the A-I$_{Milano}$ dimer has been developed recently in mice. These animals were prepared by knocking out mouse apoA-I and overexpressing either A-I$_{Milano}$ or wild type apoA-I. Mice transgenic for human apoA-II were mated with animals transgenic for apoA-I$_M$. In this way it was possible to develop double transgenic animals for apoA-I$_M$ and human apoA-II (Chiesa et al. 1998). These animals are homozygous for the mutation (differently from humans) and carry in the circulation homodimers A-I$_M$/A-I$_M$ and heterodimers A-I$_M$/A-II. Interestingly, these animals show characteristic hypertriglyceridemia mainly localized in VLDL, as well as reduced HDL levels (Chiesa et al. 1998).

Clinical Studies of sHDL with Recombinant or Extractive apoA-I

In the first clinical study with sHDL made with recombinant pro-apoA-I, four patients (two of whom had pre-existing coronary disease), all characterized by very low HDL-cholesterol levels, received a single i.v. injection of sHDL containing recombinant pro-apoA-I and soybean PC (Carlson 1995). Results were remarkable. Administration proved easy: only 10 minutes were necessary to administer 1.6 g of pro-apoA-I, side effects were apparently minimal, and antibody production could not be detected. However, the most remarkable finding in these patients, all characterized by an apparent RCT deficiency, was that HDL-cholesterol levels increased rapidly after administration of sHDL, staying elevated for at least three days after injection. In a subsequent study with the same sHDL preparation, a single infusion of sHDL (4 g of pro-apo A-I) into patients with familial hypercholesterolemia markedly increased total fecal steroid excretion (Lacko and Hiller 1997), suggesting an enhanced mobilization of peripheral cholesterol via RCT.

Two other clinical studies have investigated the effects of sHDL containing plasma-derived apoA-I during human endotoxemia, a potential target for a clinical use of the lipoprotein (Pajkrt et al. 1996, 1997). sHDL were infused into eight healthy volunteers (4-hr infusion of 40 mg/kg of apoA-I), starting 3.5 hr before an endotoxin challenge, according to a double-blind, placebo-controlled, crossover protocol. sHDL displayed a potent LPS-neutralizing activity, as indicated by the marked inhibition of endotoxin-induced release of inflammatory cytokines,

TNF, IL-6 and IL-8, and little effects on proinflammatory cytokine inhibitors IL-1ra, soluble TNF receptors and IL-10 (Pajkrt et al. 1996). Administration of sHDL also reduced the LPS-induced activation of coagulation and fibrinolysis (Pajkrt et al. 1997). These results confirm previous findings in animals (Hubsch et al. 1995; Casas et al. 1995, 1996), and clearly support the possible therapeutic use of sHDL in the treatment of septic shock. Their relevance in atherosclerosis prevention and treatment is difficult to establish, although an inhibition of the proinflammatory state that characterizes the earliest events in atherogenesis (Ross 1993) may well have important therapeutic implications.

Potential Clinical Use of sHDL with Recombinant A-I or A-I$_M$/A-I$_M$

The use of recombinant or extractive apolipoproteins in the treatment of cardiovascular diseases will, to a large extent, be dependent upon technological advances in arterial lesion evaluation. Results from the large clinical studies on plasma cholesterol reduction for the prevention of coronary disease (Scandinavian Simvastatin Survival Study Group 1994; Shepherd et al. 1995) show that a reduction of coronary risk of about 30 % can be achieved with cholesterol-lowering drug treatment. This result is likely to be related to the "stabilization" of soft, lipid-rich arterial plaques (MacIsaac et al. 1993). A direct evaluation of this hypothetical mechanism in the clinic, i.e., by direct monitoring of the coronary plaque, by technologies such as intravascular ultrasound (IVUS) has, however, generally indicated a relatively weak effect. Studies in statin-treated individuals undergoing repeated IVUS examination showed that, at the three-year interval, the unstable lipid-rich plaque volume had been reduced to a minimal extent (Takagi et al. 1997). From similarly monitored coronary arteries from transplanted hearts after four years of statin versus placebo treatment, there was an indication of a mildly reduced progression of lipid-rich lesions (Wenke et al. 1997).

   If the objective of stabilization or, even better, reduction of plaque size can be achieved by the use of repeated administrations of recombinant products, patients and doctors will be offered a much simpler approach to coronary prevention. This might turn out to be a more important target versus the classical one of preventing restenosis after angioplasty (Bauters et al. 1997). While noninvasive arterial evaluation is now a well-established procedure for monitoring the intima-media thickening in, e.g., carotid arteries (Pignoli et al. 1987), this type of approach is, as yet, poorly applicable to coronaries. On the other hand, encouraging results have been presented in both echographic (Faletra et al. 1995) and breath-hold NMR (Achenbach et al. 1997) approaches. In any case, entry of recombinant apolipoproteins into therapy might dramatically change our philosophy toward arterial disease. The concept of atherosclerosis as a disease of "plenty" will be turned around to a "deficiency" disease.

Conclusions

Discovery of the gene encoding the apoA-I$_{Milano}$ mutant was an exciting event, being the first report of a structural mutation in a human apolipoprotein. While more than 40 mutations in apoA-I have since been described, this discovery remains one of the most interesting, particularly because of the association with reduced HDL cholesterolemia and apparently increased protection from cardiovascular disease. A number of explanations have been provided for this unusual association but, as yet, a final answer has not been given. The latest, most up-to-date hypothesis links the protection exerted by the apoA-I$_{Milano}$ mutation to a relative elevation of plasma paraoxonase, an enzyme with potent antioxidant activity, carried by HDL in plasma (James et al. 1998). In contrast to other HDL mutants, where paraoxonase levels are reduced concomitant to lower HDL levels, in fact, HDL containing A-I$_{Milano}$ maintains its normal endowment of paraoxonase.

A limit to a full understanding of the potential protective effect of the A-I$_{Milano}$ is related to the fact that, up to now, no homozygous carrier has been identified. This is easily explained by the small size of the Limone population and by the rarity of the phenomenon. The observation of generally good health in A-I$_M$ homozygous transgenic mice suggests that such an event would probably not result in a lethal mutation.

The final answer as to the real protective potential of A-I$_M$ (particularly of the predominant dimeric form, with its prolonged permanence in plasma) will be provided by a wider availability of sHDL carrying recombinant A-I$_M$/A-I$_M$. This synthetic lipoprotein has shown significant potential therapeutic effects in animal models, but only in the human will final, fully reliable data be acquired. In conclusion, however, it appears that being a carrier of the A-I$_{Milano}$ mutation does, if anything, no harm and may even provide significant benefit.

References

Achenbach S, Kessler W, Moshage WE, Ropers D, Zink D, Kroeker R, Nitz W, Laub G, Bachmann K (1997) Visualization of the coronary arteries in three-dimensional reconstructions using respiratory gated magnetic resonance imaging. Coronary Artery Dis 8:441–448.

Allan CM, Fidge NH, Morrison JR, Kanellos J (1993) Monoclonal antibodies to human apolipoprotein AI: probing the putative receptor binding domain of apolipoprotein AI. Biochem J 290:449–455

Ameli S, Hultgardh Nilsson A, Cercek B, Shah PK, Forrester JS, Ageland H, Nilsson J (1994) Recombinant apolipoprotein A-IMilano reduces intimal thickening after balloon injury in hypercholesterolemic rabbits. Circulation 90:1935–1941

Baldassarre D, Gianfranceschi G, Pazzucconi F, Sirtori CR (1995) Non-invasive assessment of unstimulated forearm arterial compliance in human subjects. Impaired vasoreactivity in hypercholesterolaemia. Eur J Clin Invest 25:859–866

Bauters C, van Belle E, Meurice T, Letourneau T, Lablanche J-M, Bertrand ME (1997) Prevention of restenosis. Future directions. Trends Cardiovasc Med 7:90–94

Brewer HB, Jr., Fairwell T, Larue A, Ronan R, Houser A, Bronzert TJ (1978) The amino acid sequence of human apoA-I, an apolipoprotein isolated from high density lipoprotein. Biochem Biophys Res Commun 80:623–630

Calabresi L, Vecchio G, Longhi R, Gianazza E, Palm G, Wadensten H, Hammarstrom A, Olsson A, Karlstrom A, Sejlitz T, Ageland H, Sirtori CR, Franceschini G (1994) Molecular characterization of native and recombinant apolipoprotein A-I$_{Milano}$ dimer. The introduction of an interchain disulfide bridge remarkably alters the physicochemical properties of apolipoprotein A-I. J Bio Chem 269:32168–32174

Calabresi L, Vecchio G, Frigerio F, Vavassori L, Sirtori CR, Franceschini G (1997a) Reconstituted high-density lipoproteins with a disulfide-linked apolipoprotein A-I dimer: evidence for restricted particle size heterogeneity. Biochemistry 36:12428–12433

Calabresi L, Franceschini G, Sirtori CR, de Palma A, Saresella M, Ferrante P, Taramelli D (1997b) Inhibition of VCAM-1 expression in endothelial cells by reconstituted high density lipoproteins. Biochem Biophys Res Commun 238:61–65

Carlson LA (1995) Effect of a single infusion of recombinant human proapolipoprotein A-I liposomes (synthetic HDL) on plasma lipoproteins in patients with low high density lipoprotein cholesterol. Nutr Metab Cardiovasc Dis 5:85–91

Casas AT, Hubsch AP, Rogers BC, Doran JE (1995) Reconstituted high-density lipoprotein reduces LPS-stimulated TNF alpha. J Surg Res 59:544–552

Casas AT, Hubsch AP, Doran JE (1996) Effects of reconstituted high-density lipoprotein in persistent gram-negative bacteremia. Am Surg 62:350–355

Chiesa G, Stoltzfus LJ, Michelagnoli S, Bielicki JK, Santi M, Forte TM, Sirtori CR, Granceschini G, Rubin EM (1998) Elevated triglycerides and low HDL cholesterol in transgenic mice expressing human apolipoprotein A-I$_{Milano}$. Atherosclerosis 136:139–146

De Geest B, Zhao Z, Collen D, Holvoet P (1997) Effects of adenovirus-mediated human apo A-I gene transfer on neointima formation after endothelial denudation in apo E-deficient mice. Circulation 96:4349–4356

Duverger N, Kruth H, Emmanuel F, Caillaud JM, Viglietta C, Castro GR, Tailleux A, Fievet C, Fruchart JC, Houdebine LM, Denefle P, Castro G (1996) Inhibition of atherosclerosis development in cholesterol-fed human apolipoprotein A-I-transgenic rabbits. Circulation 94:713–717

Faletra F, Cipriani M, De Chiara F, Quattrocchi G, Danzi GB, Gronda E, Frigerio M, Mangiavacchi M, Pezzano A (1995) Imaging the left anterior descending coronary artery by high-frequency transthoracic echocardiography in heart transplant patients. Am J Cardiol 75:855–858

Forte TM, McCall MR, Amacher S, Nordhausen RW, Vigne JL, Mallory JB (1990) Physical and chemical characteristics of apolipoprotein A-I-lipid complexes produced by Chinese hamster ovary cells transfected with the human apolipoprotein A-I gene. Biochim Biophys Acta 1047:11–18

Franceschini G, Sirtori CR, Capurso A, Weisgraber KH, Mahley RW (1980) Decreased high density lipoprotein cholesterol levels with significant lipoprotein modifications and without clinical atherosclerosis in an Italian family. J Clin Invest 66:892–900

Franceschini G, Sirtori M, Gianfranceschi G, Sirtori CR (1981) Relation between the HDL apoproteins and AI isoproteins in subjects with the AI$_{Milano}$ abnormality. Metabolism 30:502–509

Franceschini G, Vecchio G, Gianfranceschi G, Magani D, Sirtori CR (1985) Apolipoprotein A-I$_{Milano}$. Accelerated binding and dissociation from lipids of a human apolipoprotein variant. J Biol Chem 260:16321–16325

Franceschini G, Calabresi L, Tosi C, Gianfranceschi G, Sirtori CR, Nichols AV (1990) Apolipoprotein A-I$_{Milano}$. Disulfide-linked dimers increase high density lipoprotein stability and hinder particle interconversion in carrier plasma. J Biol Chem 265:12224–12231

Gordon DJ, Rifkind BM (1989) High density lipoprotein: the clinical implications of recent studies. N Engl J Med 321:1311–1316

Gualandri V, Franceschini G, Sirtori CR, Gianfranceschi G, Orsini GB, Cerrone A, Menotti A (1985) AI-Milano apoprotein: identification of the complete kindred and evidence of a dominant genetic transmission. Am J Human Genet 37:1083–1097

Holvoet P, Zhao Z, Deridder E, Dhoest A, Collen D (1996) Effects of deletion of the carboxyl-terminal domain of ApoA-I or of its substitution with helices of ApoA-II on in vitro and in vivo lipoprotein association J Biol Chem 271:19395–19401

Hubsch AP, Casas AT, Doran JE (1995) Protective effects of reconstituted high-density lipoprotein in rabbit gram-negative bacteremia models. J Lab Clin Med 126:548–558

Isacchi A, Sarmientos P, Lorenzetti R, Soria M (1989) Mature apolipoprotein AI and its precursor proa-poAI: influence of the sequence at the 5' end of the gene on the efficiency of expression in *Escherichia coli*. Gene 81:129–137

James RW, Blatter Garin MC, Calabresi L, Miccoli R, von Eckardstein A, Tilly-Kiesi M, Taskinen MR, Assmann G, Franceschini G (1998) Modulated serum activities and concentrations of paraoxonase in high density lipoprotein deficiency states. Atherosclerosis 139:77–82

Jonas A (1991) Lecithin-cholesterol acyltransferase in the metabolism of high-density lipoproteins. Biochim Biophys Acta 1084:205–220

Lacko AG, Miller NE (1997) International symposium on the role of HDL in disease prevention: report on a meeting. J Lipid Res 38:1267–1273

Li D, Weng S, Yang B, Zander DS, Saldeen T, Nichols WW, Khan S, Mehta JL (1999) Inhibition of arterial thrombus formation by apo A-I$_{Milano}$. Arterioscler Thromb Vasc Biol 19:378–383

Liang HQ, Rye KA, Barter PJ (1995) Cycling of apolipoprotein A-I between lipid-associated and lipid-free pools. Biochim Biophys Acta 1257:31–37

MacIsaac AI, Thomas JD, Topol EJ (1993) Toward the quiescent coronary plaque. J Am Coll Cardiol 22:1228–1241

McGuire KA, Davidson WS, Jonas A (1996) High yield overexpression and characterization of human recombinant proapolipoprotein A-I. J Lipid Res 37:1519–1528

Mezdour H, Yamamura T, Nomura S, Yamamoto A (1995) Exogenous supply of artificial lipoproteins does not decrease susceptibility to atherosclerosis in cholesterol-fed rabbits. Atherosclerosis 113:237–246

Miyazaki A, Sakuma S, Morikawa W, Takiue T, Miake F, Terano T, Sakai M, Hakamata H, Sakamoto Y, Naito M, Ruan Y, Takahashi K, Ohta T, Horiuchi S (1995) Intravenous injection of rabbit apolipoprotein A-I inhibits the progression of atherosclerosis in cholesterol-fed rabbits. Arterioscler Thromb Vasc Biol 15:1882–1888

Moguilevsky N, Roobol C, Loriau R, Guillaume JP, Jacobs, Cravador A, Herzog A, Brouwers L, Scarso A, Gilles P (1989) Production of human recombinant proapolipoprotein A-I in *Escherichia coli*: purification and biochemical characterization. DNA 8:429–436

Moguilevsky N, Varsalona F, Guillaume JP, Gilles P, Bollen A, Roobol K (1993) Production of authentic human proapolipoprotein A-I in *Escherichia coli*: strategies for the removal of the amino-terminal methionine. J Biotechnol 27:159–172

Nanjee MN, Crouse JR, King JM, Hovorka R, Rees SE, Carson ER, Morgenthaler JJ, Miller NE, Lerch P (1996) Effects of intravenous infusion of lipid-free apo A-I in humans. Arterioscler Thromb Vasc Biol 16:1203–1214

O'Leary DH, Polak JF, Kronmal RA, Manolio TA, Burke GL, Wolfson SK (1999) Carotid-artery intima and media thickness as a risk factor for myocardial infarction and stroke in older adults. N Engl J Med 340:14–22

Oram JF, Yokoyama S (1996) Apolipoprotein-mediated removal of cellular cholesterol and phospholipids. J Lipid Res 37:2473–2491

Pajkrt D, Doran JE, Koster F, Lerch PG, Arnet B, van der Poll T, Cate JW, van Deventer SJ (1996) Antiinflammatory effects of reconstituted high-density lipoprotein during human endotoxemia. J Exp Med 184:1601–1608

Pajkrt D, Lerch PG, van der Poll T, Levi M, Illi M, Doran JE, Arnet B, Van den Ende A, ten Cate JW, van Deventer SJ (1997) Differential effects of reconstituted high-density lipoprotein on coagulation, fibrinolysis and platelet activation during human endotoxemia. Thromb Haemost 77:303–307

Paszty C, Maeda N, Verstuyft J, Rubin EM (1994) Apolipoprotein AI transgene corrects apolipoprotein E deficiency-induced atherosclerosis in mice. J Clin Invest 94:899–903

Pignoli P, Tremoli E, Poli A, Oreste P, Paoletti R (1986) Intimal plus medial thickness of the arterial wall: a direct measurement with ultrasound imaging. Circulation 74:1399–1406

Pyle LE, Sawyer WH, Fujiwara Y, Mitchell A, Fidge NH (1996) Structural and functional properties of full-length and truncated human proapolipoprotein AI expressed in *Escherichia coli*. Biochemistry 35:12046–12052

Roma P, Gregg RE, Meng MS, Ronan R, Zech LA, Franceschini G, Sirtori CR, Brewer HB Jr (1993) In vivo metabolism of a mutant form of apolipoprotein A-I, apo A-I$_{Milano}$ associated with familial hypoalphalipoproteinemia. J Clin Invest 91:1445–1452

Ross R (1993) The pathogenesis of athersclerosis: a perspective for the 1990s. Nature 362:801–809

Rothblat GH, Mahlberg FH, Johnson WJ, Philips MC (1992) Apolipoproteins, membrane cholesterol domains, and the regulation of cholesterol efflux. J Lipid Res 33:1091–1097

Rubin EM, Krauss RM, Spangler EA, Verstuyft JG, Clift SM (1991) Inhibition of early atherogenesis in transgenic mice by human apolipoprotein AI. Nature 353:265–267

Saku K, Liu R, Ohkubo K, Bai H, Hirata K, Yamamoto K, Morimoto Y, Yamada K, Arakawa K (1993) In vivo conversion of recombinant human proapolipoprotein AI (rh-Met-proapo AI) to apolipoprotein AI in rabbits. Biochim Biophys Acta 1167:257–263

Scandinavian Simvastatin Survival Study Group (1994) Randomised trial of cholesterol lowering in 4444 patients with coronary heart disease: the Scandinavian Simvastatin Survival Study (4S). Lancet 344:1383–1399

Segrest JP, Jones MK, De Loof H, Brouillette CG, Venkatachalapathi YV, Anantharamaiah GM (1992) The amphipathic helix in the exchangeable apolipoproteins: a review of secondary structure and function. J Lipid Res 33:141–166

Shah PK, Nilsson J, Kaul S, Fishbein MC, Ageland H, Hamsten A, Johansson J, Karpe F, Cercek B (1998) Effects of recombinant apolipoprotein A-I-Milano on aortic atherosclerosis in apolipoprotein E-deficient mice. Circulation 97:780–785

Shepherd J, Cobbe SM, Ford I, Isles CG, Lorimer AR, Macfarlane PW, McKillop JH, Packard CJ (1995) Prevention of coronary heart disease with pravastatin in men with hypercholesterolemia. West of Scotland Coronary Prevention Study Group. N Engl J Med 333:1301–1307

Sirtori CR, Calabresi L, Francechini G (1999) Recombinant apolipoprotein for the treatment of vascular disease. Atherosclerosis 142:29–40

Soma MR, Donetti E, Parolini C, Sirtori CR, Fumagalli R, Franceschini G (1995) Recombinant apolipoprotein A-IMilano dimer inhibits carotid intimal thickening induced by perivascular manipulation in rabbits. Circ Res 76:405–411

Sorci-Thomas MG, Parks JS, Kearns MW, Pate GN, Zhang C, Thomas MJ (1996) High level secretion of wild-type and mutant forms of human pro-apoA-I using baculovirus-mediated Sf-9 cell expression. J Lipid Res 37:673–683

Takagi T, Yoshida K, Akasaka T, Hozumi T, Morioka S, Yoshikawa J (1997) Intravascular ultrasound analysis of reduction in progression of coronary narrowing by treatment with pravastatin. Am J Cardiol 79:1673–1376

Tall A, Small DM, Shipley GG, Lees RS (1975) Apoprotein stability and lipid-protein interactions in human plasma high density lipoproteins. PNAS USA 72:4940–4942

Trachtenberg JD, Cochrane H, Sun S, Sauther M, Lassere M, Choi E, Li AP, Callow AD (1993) Apolipoprotein A-I inhibits atherosclerotic lesions progression. Circulation 88:1–552

Wald JH, Krul ES, Jonas A (1990) Structure of apolipoprotein A-I in three homogeneous, reconstituted high density lipoprotein particles. J Biol Chem 32:20037–20043

Weisgraber KH, Bersot TP, Mahley RW, Franceschini G, Sirtori CR (1980) Apoprotein A-I$_{Milano}$: isolation and characterization of a cysteine-containing variant of the A-I apoprotein from human high density lipoproteins. J Clin Invest 66:901–907

Weisgraber KH, Rall SC Jr, Bersot TP, Mahley RW, Franceschini G, Sirtori CR (1983) Apolipoprotein AI-Milano. Detection of normal AI in affected subjects: evidence for a cysteine for arginine substitution in the variant AI. J Biol Chem 258:2508–2513

Wenke K, Meiser B, Thiery J, Nagel D, von Scheidt W, Steinbeck G, Seidel D, Reichart B (1997) Simvastatin reduces graft vessel disease and mortality after heart transplantation: a four-year randomized trial. Circulation 96:1398–1402

Candidate Gene Polymorphisms in Cardiovascular Pathophysiology

F. Cambien

Candidate Genes

Candidate genes encode proteins that may play a role in the appearance and/or evolution of a disease. Until now, candidate genes were in general deduced from what is known about the biological mechanisms that may contribute to the disease. But other approaches to identifying candidate genes exist and some of them may become preeminent in the future. For example:

1) regions of interest may be deduced from linkage of random genetic markers and disease in families comprising several affected individuals. All genes within these regions become candidates;
2) in the presence of a genetic linkage in a mouse model of genetic disease, it is possible to take advantage of regions of synteny between the mouse and human genomes to identify candidate genes for the corresponding disease in humans;
3) new highly parallel technologies (cDNA arrays or DNA chips) may be applied to the identification of genes that are expressed in particular types of cells under different circumstances that are relevant to the disease;
4) ultimately, it might become possible to perform whole genome association studies, using hundreds of thousands of markers in large, population-based cohorts of patients and controls.

Polymorphism of Candidate Genes

Candidate genes are polymorphic, i.e., several forms of each gene exist that differ in term of nucleotide sequence and possibly in terms of function. In a systematic study of 36 candidate genes, we found an average of five common alleles per gene (Cambien et al. 1999). Apolipoprotein B gene is an example of a very polymorphic gene (Fig. 1). Associations among polymorphisms of a gene are frequently strong; they reflect linkage disequilibrium and result in haplotypes (combinations of alleles) that are conserved and frequently have a very ancient origin, and have therefore maintained their stability in highly variable environmental backgrounds. Their frequency as well as their impact on reproductive fitness and diverse phenotypes are likely to have changed with time, and the present situation should be viewed in the context of this long-term and highly variable

V. Boulyjenkov, K. Berg, Y. Christen (Eds.)
Genes and Resistance to Diseases
© Springer-Verlag Berlin Heidelberg 2000

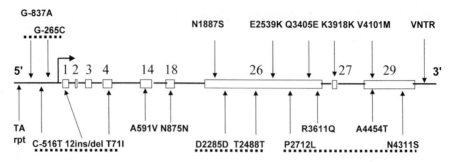

Fig. 1. Polymorphism of the apolipoprotein B gene. Most polymorphisms are shown. For the complete set of polymorphisms, and their frequencies and linkage disequilibrium, see our Internet site http://geneCanvas.idF.inserm.fr). Coding sequences are shown as white boxes, with the exon number shown above. Polymorphisms underlined by dots are in complete association (they generate only two main haplotypes). For example, G/A and G/C substitutions located in the 5' flanking region (at positions –837 and –265, respectively) were in complete association and generated only two haplotype, G^{-837}–G^{-265} and A^{-837}–C^{-265}. The C/T and A/G substitutions affecting codons 2712 and 4311, respectively, were in nearly complete association since less than 2% of subjects carried a haplotype different from the two major haplotypes, C^{C2712}–A^{C4311} and T^{C2712}–G^{C4311}. Almost complete association was also observed between the polymorphisms of the apoB gene affecting codons 2285 and 2488. Such a pattern of (nearly) complete association can extend over more than two sites. An example in the apoB gene involves the polymorphisms at position –516 from the tsp and those affecting codons 12 and 71.

genotype-phenotype relationship. Figure 2 provides a hypothetical model explaining the "survival of haplotypes." This representation helps to explain why complete association between polymorphisms of a single gene may exist and why single markers (SNPs) may be quite efficient in characterizing the variability of a gene.

Polymorphisms of Candidate Genes and Common Cardiovascular Diseases

Associations between several frequent polymorphisms of candidate genes and coronary heart disease (CHD) or its risk factors have been reported in recent years. These genes code for the apolipoproteins APOB, Moreel et al. 1992; APO-CIII, Surguchov et al. 1996; APO(a), Kraft et al. 1996; APOE, Wilson et al. 1996), lipoprotein lipase (LPL; Reymer et al. 1995); cholesterol ester transfer protein (CETP, Fumeron et al. 1995), fibrinogen (Fowkes et al. 1992), plasminogen activator inhibitor type 1 (PAI-1, Eriksson et al. 1995), angiotensin I converting enzyme (ACE, Cambien et al. 1992), angiotensin II receptor (AT1R, Tiret et al. 1994), angiotensinogen (Katsuya et al. 1995), paraoxonase (Ruiz et al. 1995), methylene tetrahydrofolate reductase (MTHFR; Frosst et al. 1995), E-selectin (Wenzel et al. 1994), P-selectin (Hermann et al. 1998), glycoprotein GpIIIa receptor subunit (Weiss et al. 1996), Transforming growth factor β (TGFβ; Cambien et al. 1996) and coagulation factor XIII (Kohler et al. 1998). Some of these polymorphisms have been extensively studied and the pathophysiological relevance of some of them is debatable. It is likely that a number of reported associations will be invalidated in

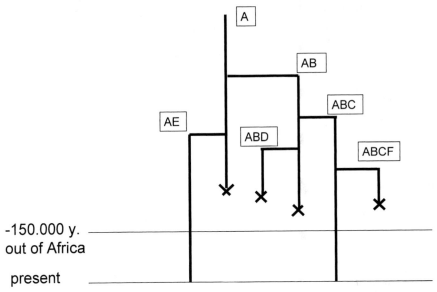

Fig. 2. A hypothetical example of two haplotypes having survived a complex history. Only haplotypes AE and ABC have survived to the present time. They differ at three sites: E, B and C. Note that variants D and F existed in the past but have now disappeared. Polymorphism loss may explain part of the relatively low nucleotide diversity that has been reported in humans. Lost haplotypes may be viewed as a negative trace of the hazards and conflicts that have punctuated human evolution (realising this may be source of some emotion, as in an assembly where an empty chair reminds somebody of a deceased friend). The genetic polymorphism of humans at present is only part of the polymorphism accumulated by the human species over time. Although a number of haplotypes may be lacking in some present human populations as a consequence of genetic drift, common haplotypes are likely to be found in most populations, but possibly with different frequencies. A repeated aggregation-dispersion process may adequately represent the formation of the human groups in time and place and may account for the different distributions of human genes in different populations. Superficially, it might appear that such representation place great emphasis on chance as the main underlying process explaining the genetic structure of the human population at any point in time. However, assuming that human groups aggregate and get dispersed randomly may be naive.

the future. There is also little doubt that other genes contributing to CHD and other common cardiovascular disorders (CVD) will be identified.

Gene Polymorphisms are Response Modifiers

What is known about the epidemiology and pathophysiology of common CVD suggests that the genetic component of these disorders involves polymorphic genes that act as response modifiers to environmental and pathophysiological challenges. The most realistic genetic model for common CVD assumes : 1) several genes with 2) one or several common polymorphic loci that 3) have a modest effect on disease risk and 4) may interact with each other and with nongenetic factors. We are well aware of the importance of environment on CVD epidemiol-

ogy. Environmental factors are likely to explain most of the differences of prevalence across different populations, time-related changes in incidence within populations and changes in frequency of the disease in migrant populations. Under
ideal environmental conditions (low intake of saturated fats and salt, appropriate
level of physical activity and absence of smoking), hypertension, atherosclerosis
and CVD in general would probably be rare. It appears paradoxical therefore that
genetic factors play an important role in susceptibility to these disorders, as
attested to by the results of numerous studies (Marenberg et al. 1994). The resolu-
tion of this paradox is simple: genetic and nongenetic factors interact or, in other
words, the genetic variants that have an impact on function act as response modifiers of the effect of environmental factors. In research practice, this somewhat
obvious principle is frequently forgotten, as the inappropriateness of commonly
used study designs indicates.

It is a challenge for future studies to assemble as much information as possible that may be relevant to investigate gene-gene and gene-nongene interactions
and to subcategorise the patients. It is also important to realise that considering
polymorphic genes as putative response modifiers may reveal their ambivalent
nature with regard to present and past phenotypes.

Some Examples of Gene Polymorphisms Acting as Response Modifiers

Examples showing that candidate gene polymorphisms are response modifiers
accumulate rapidly: the polymorphism of the MTHFR gene modulates the effect
of folate intake on plasma folate (Jaques et al. 1996; Harmon et al. 1996); polymorphisms of the CETP gene affect the relationship between alcohol intake and
plasma HDL cholesterol (Fumeron et al. 1995); polymorphisms of the beta fibrinogen gene affect the relationship between cigarette smoking or intense physical
activity and plasma fibrinogen (Humphries et al. 1999); the αAdducin polymorphism appears to affect the effect of salt intake on blood pressure (Cusi et al.
1997); the ACE polymorphism, which explains a large fraction of the variability
of plasma and cellular ACE, modifies the effect of physical training on left ventricular mass, endurance and energy balance (Montgomery et al. 1999); the same
polymorphism influences the consequence of Stent angioplasty on restenosis
(Amant et al. 1997; Ribichini et al. 1998), etc. Environmental factors are not the
only challenging factors; genes may also modulate the response to pathophysiological triggers. For example, hypertension is known to be associated with
increased large artery wall stiffness but differently, according to a polymorphism
of the angiotensin AT1 receptor (Benetos et al. 1996); the ACE polymorphism
accelerates the degradation of renal function in patients with diabetic or Iga
nephropathy (Yoshida et al. 1996) and of congestive heart failure postmyocardial
infarction (Pinto et al. 1995); a polymorphism of the beta fibrinogen gene may be
associated risk of myocardial infarction in patients with severe atherosclerosis
(Behague et al. 1996) etc.

Genotype-Phenotype Associations May Be Complex

We have seen above that, as a consequence of tight association among polymorphisms within gene, single markers may be quite efficient in characterizing the variability of that gene. However in many cases the situation may not be so simple. For example, studies aimed at dissecting the genetic component of plasma ACE levels determined the existence of a major gene effect that would be attributable to an unidentified functional polymorphism within the ACE gene, in complete linkage disequilibrium with an ACE insertion/deletion polymorphism located in intron 16 (Tiret et al. 1992). To characterize this functional polymorphism, an extensive molecular screening of the 5', coding and 3' sequences of the ACE gene was performed, which identified 10 new polymorphisms generating three main haplotypes. Analysis of the relationship of these polymorphisms to plasma ACE levels revealed that there were probably two functional polymorphisms in strong linkage disequilibrium with each other, and acting additively on ACE levels (Vilard et al. 1996). Another example is provided by the CETP gene, in which a *Taq*1 polymorphism was shown to be associated with both CETP mass and HDL-cholesterol, in interaction with alcohol consumption (Fumeron et al. 1995). A systematic molecular screening of the CETP gene further suggested that there were at least three functional polymorphisms influencing CETP mass and HDL-cholesterol levels through distinct mechanisms (Corbex et al. 2000). These examples suggest that different genotypes of a gene may affect different phenotypes and that beneficial as well as deleterious effects may be expected.

Bad or Good Genes

Asserting that a genotype is beneficial or deleterious is relative to the choice of a phenotype measured in a population at a particular time and place. In doing so, we ignore all other known or unknown phenotypes that are related to the genotype, including those that may affect reproductive fitness and might, therefore, account for the prevalence of particular alleles. We also ignore possible heterogeneity across populations. A typical question that may be asked is, therefore, "Is this allele associated with an increased risk of disease A in population X at time T?" A response to such a question may be of some practical value, but it is so contextual that its scientific validity is questionable (we are not allowed to extrapolate this response to population X' at time T'). To scientifically address the relationship between a functional polymorphism (i.e., a polymorphism that is associated with difference in one or several biological functions) and phenotypes implies asking a number of interconnected questions:

1) How are the biological functions affected? Is it possible to evaluate this by cellular models?
2) How is physiology affected by the polymorphism? To evaluate this we need appropriate measures for investigating relevant systems in humans. This research may also benefit from animal models.

3) What is the influence of the polymorphism on the different disease endpoints that may result from the impact of the polymorphism on biological and physiological functions?

4) What is the relationship of the polymorphism with survival? Since its impact may change with age (Toupance et al. 1998), age-related survival and age-related frequencies of each genotype should be evaluated.

5) All questions above should also be asked in the context of possible interactions between the polymorphism and other variable individual characteristics, genetic and nongenetic.

Conclusion

Future – The Genetic Side

Any plan for future studies cannot ignore the potential that will be provided by the completion of the Human Genome Project and by the new technologies that have been developed for high throughput sequencing, polymorphism detection and genotyping. Using these new tools we plan to investigate 500 candidate genes during the next three to four years. The list of candidate genes has been established from the known biological role of their product; however, they may represent only a small fraction of all candidate genes. Available technology allows the identification of new candidates by comparing gene expression in two cell lines that differ for a particular phenotype of interest. Micro-arrays and DNA chips have a considerable potential for generating differential expression profiles that, coupled with genomic sequences, will allow rapid and massive identification of new candidate genes.

Future – The Phenotypic Side

More phenotypes and more accurate phenotypes are required. This will be the most difficult aspect of future studies. It is neglected for the moment because the genetic side is attracting most of the attention and money; however, the naïveté of an approach focussed mainly on the genotype will rapidly become obvious. mRNA, cellular and circulating proteins, cellular and physiological functions, preclinical measurements and well-defined clinical sub-entities will be required for assessing the importance of candidate gene polymorphisms. To correctly assess these phenotypes, dense networks of collaborating laboratories will have to be developed.

References

Amant C, Bauters C, Bodart JC, Lablanche JM, Grollier G, Danchin N, Hamon M, Richard F, Helbecque N, McFadden EP, Amouyel P, Bertrand ME (1997) D allele of the angiotensin I-converting enzyme is a major risk factor for restenosis after coronary stenting. Circulation 96:56–60

Behague I, Poirier O, Nicaud V, Evans A, Arveiler D, Luc G, Cambou JP, Scarabin PY, Para L, Green F, Cambien F (1996) Beta fibrinogen gene polymorphisms are associated with plasma fibrinogen and coronary artery disease in patients with myocardial infarction – The ECTIM Stud. Circulation 93:440–449

Benetos A, Gautier S, Ricard S, Topouchian J, Asmar R, Poirier O, Larosa E, Guize L, Safar M, Soubrier F, Cambien F (1996) Influence of angiotensin converting enzyme and angiotensin II type 1 receptor gene polymorphisms on aortic stiffness in normotensive and hypertensive patients. Circulation 94:698–703

Cambien F, Poirier O, Lecerf L, Evans A, Cambou J-P, Arveiler D, Luc G, Bard J-M, Bara L, Ricard S, Tiret L, Amouyel P, Alhenc-Gelas F, Soubrier F (1992) Deletion polymorphism at the angiotensin-converting enzyme gene is a potent risk factor for myocardial infarction. Nature 359:641–644

Cambien F, Poirier O, Nicaud V, Herrmann SM, Mallet C, Ricarol S, Behague I, Hallet V, Blanc H, Lonkaci V, Thillet J, Evans A, Ruidavets JB, Arveiler D, Luc G, Tirel L (1999) Sequence diversity in 36 Candidate genes for Cardiovascular disorders. Am J Hum Genet 65:183–191

Cambien F, Ricard S, Troesch A, Mallet C, Generenaz L, Evans A, Arveiler D, Luc G, Ruidavets JB, Poirier O (1996) Polymorphisms of the Transforming Growth Factor β1 Gene in relation to myocardial infarction and blood pressure, the ECTIM Study. Hypertension 28:881–887

Corbex M, Poirier O, Tumeron F, Betoulle D, Evans A, Ruidavets JB, Arveiler D, Luc G, Tiret L, Cambien F (2000) Extensive association analysis between CETP gene and coronary heart disease phenotypes reveal several putative functional polymorphisms and gene-environment interaction. Genet Epidemiol, in press

Cusi D, Barlassina C, Azzani T, Casari G, Citterio L, Devoto M, Glorioso N, Lanzani C, Manunta P, Righetti M, Rivera R, Stella P, Troffa C, Zagato L, Bianchi G (1997) Polymorphisms of alpha-adducin and salt sensitivity in patients with essential hypertension. Lancet 349:1353–1357

Eriksson P, Kallin B, van't Hooft FM, Bavenholm P, Hamsten A (1995) Allele-specific increase in basal transcription of the plasminogen-activator inhibitor 1 gene is associated with myocardial infarction. Proc Natl Acad Sci USA 92:1851–1855

Fowkes FGR, Connor JM, Smith FB, Wood J, Donnan PT, Lowe GD (1992) Fibrinogen genotype and risk of peripheral atherosclerosis. Lancet 339:693–696

Frosst P, Blom HJ, Milos R, Goyette P, Sheppard CA, Matthews RG, Boers GJH, den Heijer M, Kluijtmans LAJ, van den Heuvel LP, Rozen R (1995) A candidate genetic risk factor for vascular disease: a common mutation in methylenetetrahydrofolate reductase. Nat Genet 10:111–113

Fumeron F, Betoulle D, Luc G, Behague I, Ricard S, Poirier O, Riadh Jemaa, Evans A, Arveiler D, Marques-Vidal P, Bard J-M, Fruchart J-C, Ducimetière P, Apfelbaum M, Cambien F (1995) Alcohol intake modulates the effect of a polymorphism of the cholesteryl ester transfer protein gene on plasma high density lipoprotein and the risk of myocardial infarction. J Clin Invest 96:1664–1671

Harmon DL, Woodside JV, Yarnell JW, McMaster D, Young IS, McCrum EE, Gey KF, Whitehead AS, Evans AE (1996) The common "thermolabile" variant of methylene tetrahydrofolate reductase is a major determinant of mild hyperhomocysteinaemia. Q J Med 89:571–577

Herrmann SF, Ricard S, Nicaud V, Mallet C, Evans A, Ruidavets JB, Arveiler D, Luc G, Cambien F (1998) The P-selectin gene is highly polymorphic: reduced frequency of the PRO715 allele carriers in patients with myocardial infarction. Human Mol Genet 7:1277–1284

Humphries SE, Henry JA, Montgomery (1999) Gene-environment interaction in the determination of levels of haemostatic variables involved in thrombosis and fibrinolysis. Blood Coagul Fibrinolysis 10 Suppl:S17–21.

Jacques PF, Bostom AG, Williams RR, Ellison RC, Eckfeldt JH, Rosenberg IH, Selhub J, Rozen R (1996) Relation between folate status, a common mutation in methylene-tetrahydrofolate reductase, and plasma homocysteine concentrations. Circulation 93:7–9

Katsuya T, Koike G, Yee TW, Sharpe N, Jackson R, Norton R, Horiuchi M, Pratt RE, Dzau VJ, MacMahon S (1995) Association of angiotensinogen gene T235 variant with increased risk of coronary heart disease. Lancet 345:1600–1603

Kohler HP, Stickland MH, Ossei-Gerning N, Carter A, Mikkola H, Grant PJ (1998) Association of a common polymorphism in the factor XIII gene with myocardial infarction. Thromb Haemost 79:8–13

Kraft HG, Lingenhel A, Kochl S, Hoppichler F, Kronenberg F, Abe A, Muhlberger V, Schonitzer D, Utermann G (1996) Apolipoprotein(a) kringle IV repeat number predicts risk for coronary heart disease. Arterioscler Thromb Vasc Biol 16:713–719

Marenberg ME, Risch N, Berkman LF, Floderus B, de Faire U (1994) Genetic susceptibility to death from coronary heart disease in a study of twins. New Engl J Med 330:1041–1046

Montgomery H, Clarkson P, Barnard M, Bell J, Brynes A, Dollery C, Hajnal J, Hemingway H, Mercer D, Jarman P, Marshall R, Prasad K, Rayson M, Saeed N, Talmud P, Thomas L, Jubb M, World M, Humphries S (1999) Angiotensin-converting-enzyme gene insertion/deletion polymorphism and response to physical training. Lancet 353:541–545

Moreel JFR, Roizes G, Evans AE, Arveiler D, Cambou JP, Souriau C, Parra HJ, Desmarais E, Fruchart JC, Ducimetière P, Cambien F (1992) The polymorphism ApoB/4311 in patients with MI and controls: the ECTIM Study, Human Genet 89:169–175

Pinto YM, van Gilst WH, Kingma JH, Schunkert H (1995) Deletion-allele of the angiotensin-converting enzyme gene is associated with progressive ventricular dilatation after anterior myocardial infarction. J Am Coll Cardiol 25:1622–1626

Reymer PWA, Gagné E, Groenemeyer BE, Zhang H, Forsyth I, Jansen H, Seidell JC, Kromhout D, Lie KE, Kastelein J, Hayden MR (1995) A lipoprotein lipase mutation (Asn291Ser) is associated with reduced HDL cholesterol levels in premature atherosclerosis. Nat Genet 10:28–34

Ribichini F, Steffenino G, Dellavalle A, Matullo G, Colajanni E, Camilla T, Vado A, Benetton G, Uslenghi E, Piazza A (1998) Plasma activity and insertion/deletion polymorphism of angiotensin I-converting enzyme – a major risk factor and a marker of risk of coronary stent restenosis. Circulation 97:147–154

Ruiz J, Blanche H, James RW, Garin MC, Vaisse C, Charpentier G, Cohen N, Morabia A, Passa P, Froguel P (1995) Gln-Arg192 polymorphism of paraoxonase and coronary heart disease in type 2 diabetes. Lancet 346:869–872

Surguchov AP, Page GP, Smith L, Patsch W, Boerwinkle E (1996) Polymorphic markers in apolipoprotein CIII gene flanking regions and hypertriglyceridemia. Arterioscler Thromb Vasc Biol 16:941–947

Tiret L, Rigat B, Visvikis S, Breda C, Corvol P, Cambien F, Soubrier F (1992) Evidence from combined segregation and linkage analysis, that a variant of the Angiotensine I – converting enzyme (ACE) gene controls plasma ACE levels. Am J Human Gen 51:197–205

Tiret L, Bonnardeaux A, Poirier O, Ricard S, Marques-Vidal P, Evans A, Arveiler D, Luc G, Kee F, Ducimetière P, Soubrier F, Cambien F (1994) Synergistic effects of angiotensin-converting enzyme and angiotensin-II type 1 receptor gene polymorphisms on the risk of myocardial infarction. Lancet 344:910–913

Toupance B, Godelle B, Gouyon PH, Schachter F (1998) A model for antagonistic pleiotropic gene action for mortality and advanced age. Am J Human Genet 62:1525–1534

Vilard E, Tiret L, Visvikis S, Rakotovao R, Cambien F, Soubrier F (1996) Identification of new polymorphisms of the angiotensin I-converting enzyme (ACE) gene, and study of their relationship to plasma ACE levels with two-QTL segregation-linkage analysis. Am J Human Genet 58:1268–1278

Weiss EJ, Bray PF, Tayback M, Schulman SP, Kickler TS, Becker LC, Weiss JL, Gerstenblith G, Goldschmidt-Clermont PJ (1996) A polymorphism of a platelet glycoprotein receptor as an inherited risk factor for coronary thrombosis. New Engl J Med 334:1090–1094

Wenzel K, Felix S, Kleber FX, Brachold R, Menke T, Schattke S, Schulte KL, Gläser C, Rohde K, Baumann G, Speer A (1994) E-selectin polymorphism and atherosclerosis: an association study. Hum Mol Genet 3:1935–1937

Wilson PWF, Schaefer EJ, Larson MG, Ordovas JM (1996) Apolipoprotein E alleles and risk of coronary disease – a meta analysis, Arterioscler. Thromb Vasc Biol 16:1250–1255

Yoshida H, Kon V, Ichikawa I (1996) Polymorphisms of the renin-angiotensin system genes in progressive renal diseases. Kidney Int 50:732–744

Protective Responses of Endothelial Cells

M. P. Soares, C. Ferran, K. Sato, K. Takigami, J. Anrather, Y. Lin and F. H. Bach

Summary

Endothelial cells (EC) as they normally exist in their quiescent state perform critical functions in maintaining blood flow and avoiding thrombosis. Various pro-inflammatory stimuli can induce EC to be activated, which results in recruitment, trans-endothelial migration and activation of circulating leukocytes, pro-coagulation, platelet aggregation, and other responses associated with inflammation. In the case of an organ that is transplanted, these reactions associated with EC activation accompany the rejection of such organ. We suggested, several years ago, that EC activation is the underlying basis of rejection of organ xenografts, i.e., grafts such as a heart or kidney transplanted across different species. While antibodies and complement in the recipient are clearly implicated in EC activation and xenograft rejection, investigators in the 1980s showed that, under certain circumstances, grafts can survive indefinitely despite the presence of anti-graft antibodies and complement. We referred to the survival of an organ in the presence of anti-organ antibodies and complement as "accommodation." One possible mechanism that we proposed to explain accommodation of these grafts was that, under certain circumstances the EC in the graft up-regulate the expression of "protective genes" that would prevent those reactions associated with EC activation that presumably lead to rejection. We have since found that such protective genes do exist and that they can play such a role.

Introduction

Rejection of a xenograft, i.e., an organ transplanted from one species to another, can be used as a model to study endothelial cell (EC) biology. From the clinical perspective, it is hoped that we will be able to transplant organs from pigs to humans. As pre-clinical models, we use rats as recipients of hearts or other organs from mice, hamsters or guinea pigs, as well as transplanting pig organs to non-human primates.

We hypothesized some time ago that xenograft rejection was based on EC activation (Bach et al. 1994). In fact, rejection of an organ transplanted from one species to another does involve EC activation (Blakely et al. 1994; Lesnikoski et al. 1995). While there is still no definitive evidence for this hypothesis, all of the data

V. Boulyjenkov, K. Berg, Y. Christen (Eds.)
Genes and Resistance to Diseases
© Springer-Verlag Berlin Heidelberg 2000

others and we have collected are consistent with such a model. Based on our studies of EC in xenograft rejection, we found that in addition to the classical responses associated with EC activation that have been described over the last two decades activated EC under certain circumstances up-regulate genes that we refer to as "protective genes" (Bach et al. 1997a). We believe that expression of these protective genes during EC activation represents a general regulatory response in EC that occurs physiologically and probably takes place in other cells as well. We regard this as a response that functions both as a negative feedback reaction to block unfettered activation of the cell as well as a mechanism cells use to avoid their own death. Unfettered activation in this model is hypothesized to contribute to cell death (Bach et al. 1997a).

The protective genes that others and we have identified in each case are anti-apoptotic genes (Bach et al. 1997a). In particular, the genes that we have studied all have the additional function of suppressing the activation response of the cells in which they are expressed (Bach et al. 1997a). As mentioned earlier, we interpret this not only as a negative feedback reaction but also as a mechanism used by the cell to avoid excessive responses that, a side effect, evoke apoptosis of the cell.

Xenograft Rejection and Endothelial Cell Activation

Until recently the major obstacle to the clinical applications of xenotransplantation was that immediately vascularized porcine organs were inexorably rejected in a few minutes or hours when transplanted into non-human primates (Lambrigts et al. 1998). The mechanism of rejection, referred to as hyperacute rejection (HAR), is similar to that of immediately vascularized allografts transplanted across incompatible ABO-blood groups (Slapak et al. 1981; Alexandre et al. 1987; Bannett et al. 1987). The pathogenesis of HAR is always associated with the presence in the recipient of high serum levels of preformed natural antibodies that recognize one or several antigens expressed on the endothelium of the graft. In the case of ABO incompatible allografts, these preformed antibodies recognize the blood group antigens A or B (Latinne et al. 1989). In porcine xenografts, these blood group antigens are not expressed, but xenografts express a similar carbohydrate molecule, [Gal-α(1,3)-Gal-β(1,4)-GlcNac], referred to as α-Gal (Good et al. 1992; Galili 1993). Humans do not express α-Gal (Galili and Swanson 1991) and therefore generate natural antibodies that recognize this epitope as a part of their pre-immune repertoire (Galili et al. 1984, 1985). In addition, α-Gal is expressed in bacteria of the intestinal flora (Galili et al. 1998) and thus immunization against bacterial antigens containing this epitope is likely to contribute to the high serum levels of anti-α-Gal antibodies (0.1 µg/ml IgM; 1 mg/ml IgG) in humans (Galili et al. 1984, 1988; Vanhove and Back 1993). The pathogenesis of HAR of pig to primate xenografts is initiated by the binding of anti-α-Gal antibodies to the xenograft endothelium (Collins et al. 1995). Upon cross-linking on the surface of xenograft EC, anti-α-Gal antibodies of the IgM isotype activate the classical pathway of complement which leads to EC activation (Saadi and Platt 1995; Tedesco et al. 1997; van den Berg et al. 1998; Lozada et al. 1995). While in their quiescent state, EC prevent coagulation and platelet thrombosis; these functions can be lost during EC activation. The first stage of EC activation referred to as type I EC activation occurs in a few minutes, independently of gene transcription or protein synthesis (Bach et al. 1994; Cotran and Pober 1989; Pober and Cotran 1990). Type I EC activation is characterized by the translocation to the EC surface of the adhesion molecule P-selectin (Hattori et al. 1989a, b), the release of pro-coagulant Von Willebrand factor (Reinders et al. 1998), the secretion of the chemokine IL-8 (Utgaard et al. 1998; Wolff et al. 1998) and shedding from the EC surface of several anti-thrombotic molecules, including, 1) heparan sulfate/anti-thrombin III (Platt et al. 1990), 2) thrombomodulin (Scarpati and Sadler 1989) and 3) ATPDase/CD39 (Robson et al. 1997). In addition, EC undergo transient retraction from one another with exposure of the pro-thrombotic sub-endothelium (Saadi and Platt 1995). Presumably these phenotypic modifications play a central role in the pathogenesis of HAR, since the endothelium can transform in few minutes from its normally anti-thrombotic state into a highly pro-thrombotic one that promotes the thrombosis and hemorrhage that characterize HAR (Bach et al. 1996).

Given the similarities between HAR of pig- to -primate xenografts and that of ABO-incompatible allografts, the expectation was that if HAR of xenografts could be overcome, the only remaining problem would be the xenograft equivalent of T cell-mediated rejection of allografts. Unfortunately, this did not eventuate (Bach et al. 1996; Platt 1998; Parker et al. 1996). HAR of pig- to -primate xenografts can be overcome either by depleting preformed anti-α-Gal antibodies from the xenograft recipient (Xu et al. 1998; Lin et al. 1998) or by blocking complement activation (Pruitt et al. 1997; Kobayashi et al. 1997; Leventhal et al. 1994). However, xenografts are still rejected three to four days after transplantation through a process referred to as delayed xenograft rejection (DXR; Bach et al. 1996; Soares et al.). The pathogenesis of DXR is probably initiated by the rapid generation of elicited anti-graft antibodies (Miyatake et al. 1998; Soares et al. 1999) directed mainly against the α-Gal epitope (Lin et al. 1998, 1997; McCurry et al. 1997). Because these antibodies recognize carbohydrate epitopes, including α-Gal, they can be generated in a T cell-independent manner (Candinas 1996), which probably explains why DXR occurs in the absence of T cells, e.g., under T cell immunosuppression (Lin et al. 1998; Pruitt et al. 1997; Kobayashi et al. 1997; Leventhal et al. 1994). As for preformed antibodies, elicited antibodies of the IgM isotype activate the classical pathway of complement. However, complement activation remains at sublytic levels because, in most cases, complement inhibitors are used to prevent HAR (Lin et al. 1998; Pruitt et al. 1997; Kobayashi et al. 1997; Leventhal et al. 1994). Nevertheless, membrane-bound C1q (van den Berg et al. 1998; Lozada et al. 1995) and/or sublytic C5b-9 (Tedesco et al. 1997; Saadi et al. 1995) are sufficient to induce a series of modifications of the EC phenotype that occurs in a few hours and depends on the activation of gene transcription and protein synthesis. These responses are referred to as type II EC activation (Bach et al.

1994; Cotran and Pober 1989; Pober and Cotran 1990). This later phase of EC activation is characterized by the up-regulation of a series of pro-inflammatory genes encoding the adhesion molecules E-selectin, P-selectin, ICAM-1 and VCAM-1, the cytokines/chemokines IL-1α/β, IL-6, IL-8, MCP-1, MIP-1α and RANTES, as well as the pro-coagulant molecules plasminogen activator inhibitor type 1 (PAI-1) and tissue factor (TF) (Bach et al. 1994; Cotran and Pober 1989; Pober and Cotran 1990; Mantovani et al. 1997). Presumably, the expression of these pro-inflammatory genes contributes to the pathogenesis of DXR by promoting leukocyte activation, adhesion and transmigration as well as the coagulation and platelet aggregation that characterize DXR (Bach et al. 1996, 1997b; Platt 1998).

In addition to the expression of pro-inflammatory genes, type II EC activation is also associated with the expression of several anti-inflammatory genes (Bach et al. 1997a). We refer to these genes as "protective genes" because, in vitro, they inhibit the expression of pro-inflammatory genes associated with EC activation and, in addition protect EC from undergoing apoptosis (Bach et al. 1997a). These protective genes include the zinc finger molecule A20 (Cooper et al. 1996; Opipari et al. 1992), the bcl family member A1 (Scarpati and Sadler 1989; Karsan et al. 1996), the anti-oxidant manganese superoxide dismutase (Mn-SOD; Wong et al. 1989; Stroka et al. 1999) and probably several members of the family of inhibitors of apoptosis (IAP; Roy et al. 1997; Chu et al. 1997) as well as the recently described IEX-1-L gene (Wu et al. 1998).

Sequence comparison of the regulatory regions of most of the immediate early responsive genes expressed during EC activation reveals that essentially all of them contain at least one binding site for the transcription factor NF-κB (Collins et al. 1995). More importantly, the transcription of these genes is strictly dependent on the activation of NF-κB, as illustrated by the observation that blockage of this transcription factor results in suppression of the expression of these genes (Anrather et al. 1997; Soares et al. 1998a; Wrighton et al. 1996). Other protective genes, however, such as the bcl family members bcl-2 and bcl-x_L (Bach et al. 1997a; Badrichani 1999) or the stress responsive genes hsp-70 and hsp-32 (heme oxygenase-1 (HO-1)), can be expressed in EC but are not immediate early responsive genes and are not dependent on the transcription factor NF-κB for their expression.

Our hypothesis has been that DXR results in large measure from the uncontrolled expression of pro-inflammatory genes associated with the activation of xenograft EC (Bach et al. 1997c). Based on this model, we have proposed that suppression of the expression of pro-inflammatory genes associated with EC activation may contribute to overcoming DXR (Bach et al. 1997c; Soares et al. 1999b). We have shown that the expression of these pro-inflammatory genes is efficiently suppressed once the activation of the transcription factor NF-κB is blocked (Anrather et al. 1997; Soares et al. 1998a; Wrighton et al. 1996). However, inhibition of this transcription factor can sensitize cells to undergo apoptosis (Vanantwerp et al. 1996; Wang et al. 1996; Baichwal and Baeuerle 1997; Beg and Baltimore 1996), since the expression of protective (anti-apoptotic genes is also suppressed when NF-κB activation is blocked (Soares et al. 1998a). An alternative approach would be to promote the expression of protective genes in EC since many of these genes block the activation of the transcription factor NF-$\varkappa$B while inhibiting EC apoptosis (Bach et al. 1997a; Ferran et al. 1998).

Heme Oxygenase-1 (HO-1) as a Protective Gene

Together with the constitutive form of HO, heme oxygenase-2 (HO-2), HO-1 is the rate-limiting enzyme in the catabolism of heme to yield equimolar quantities of biliverdin Ixa, iron and carbon monoxide (CO), with biliverdin being subsequently catabolyzed into bilirubin (Maines 1997; Choi and Alam 1996). The physiological role of HO-1 includes the maintenance of iron homeostatis (Poss and Tanegawa 1997a; Yachie et al. 1999) and cytoprotection against oxidative stress (Yachie et al. 1999; Poss and Tonegawa 1997b). In addition, up-regulation of HO-1 suppresses a variety of inflammatory responses in vivo including endotoxic shock (Otterbein et al. 1995; 1997), hyperoxia (Lee et al. 1996) and acute pleurisy (Willis et al. 1996), whereas inhibition of HO-1 exacerbates these inflammatory responses. Additional anti-inflammatory effects of HO-1 have been suggested to include modulation of monocyte/macrophage activation (Otterbein et al., in press) transmigration across the endothelium (Ishikawa et al. 1997) as well as inhibition of EC apoptosis (Soares et al. 1998b; Petrache et al. 1997).

Little is know about the role of HO-1 in transplantation (Soares et al. 2000). Expression of HO-1 has been reported in leukocytes infiltrating allografts (Agarwal et al. 1996). Work from our laboratory has demonstrated that long-term survival of cardiac xenografts is associated with expression of HO-1 in the xenograft endothelium and smooth muscle cells (Bach et al. 1997d; Koyamada et al. 1998). In addition, we have recently demonstrated that expression of HO-1 in these xenografts is essential to insure their long-term survival (Soares et al. 1998b). More recent data suggest that expression of HO-1 in EC and smooth muscle cells may protect grafts from chronic dysfunction (Hancock et al. 1998).

The molecular mechanism(s) responsible for the cytoprotective effects of HO-1 remains largely unknown. The current view is that HO-1 has a diverse spectrum of effects that are associated with the different end products of heme catabolism generated through HO-1 enzymatic activity, e.g., bilirubin, free iron and CO (Maines 1997; Choi and Alam 1996). Bilirubin has been shown to be a potent anti-oxidant (Stocker et al. 1987) and, as for other anti-oxidants, may inhibit the generation of reactive oxygen species, a well-established component of the signaling pathways leading to the transcriptional up-regulation of pro-inflammatory genes as well as apoptosis (Buttke and Sandstrom 1994). Degradation of heme by HO-1 results in the release of iron which has the potential to exacerbate the cytotoxic effects of reactive oxygen species (Balla et al. 1992, 1993; Platt and Nath 1998). However, generation of intra-cellular free iron by HO-1 up-regulates the expression of the iron chelator ferritin (Eisenstein et al. 1991; Harrison et al. 1996), which has a high capacity to store free iron (Picard et al. 1998). Ferritin has been shown to protect EC from activated neutrophils as well as

H_2O_2-mediated cytotoxicity (Balla et al. 1992), which suggests that some of the effects of HO-1 may be mediated indirectly by ferritin (Balla et al. 1992, 1993). Another end product of heme degradation by HO-1, the gaseous molecule CO, may have potent anti-inflammatory effects as well. As for nitric oxide (NO), CO plays an important role in regulating vasomotor tone by promoting vasorelaxation (Wagner et al. 1997; Morita et al. 1995). This may be a crucial feature responsible for the cytoprotective effects of HO-1 expression in EC and smooth muscle cells since vasorelaxation may allow maintenance of blood flow at sites of inflammation, thus countering those effects of coagulation and thrombosis that can lead to anoxia and tissue necrosis. These effects of CO seem to be mediated through the activation of guanylyl cyclase and subsequent cGMP generation upon binding of CO to the heme moiety of this enzyme (Wagner et al. 1997). CO may have additional anti-inflammatory effects, such as the ability to inhibit platelet activation/aggregation (Wagner et al. 1997).

Protective Genes and Xenograft Rejection

In apparent contradiction to the ability of antibodies and complement to precipitate graft rejection, investigators in the 1980s showed that human-to-human kidneys transplanted across an ABO-incompatible blood barrier could survive in some cases despite the presence of anti-blood group antibodies and complement that could deposit on the EC of the kidney. These investigators depleted the offending anti-AB antibodies before transplantation and maintained them at reduced levels for three to four days thereafter (Poss and Tanegawa 1997a, b; Soares et al. 1998b). Later, the antibodies returned to pre-depletion levels in the presence of normal levels of complement and yet the kidneys survived (Poss and Tanegawa 1997a, b; Soares et al. 1998b). We referred to the survival of an organ in the presence of anti-organ antibodies and complement as "accommodation" (Soares et al. 1999b). One possible mechanism that we proposed to explain accommodation was that, under certain circumstances, the EC of the transplanted organ upregulate "protective gene" (Soares et al. 1999b). Products of these protective genes would prevent those reactions associated with EC activation that would otherwise lead to rejection. We have since found that such protective genes do exist and that they can play an essential role to assure xenograft survival.

To study this phenomenon we have used as an experimental model the transplantation of mouse or hamster hearts in to rats (Soares et al. 1999b). We and others have found that a large percentage of hamster hearts transplanted into rats treated briefly with the complement inhibitor cobra venom factor (CVF) and the T cell immunosuppressive cyclosporin A (CsA) survives indefinitely, despite the presence of anti-graft antibodies and complement, i.e., they accommodated (Bach et al. 1997d; Hasan et al. 1992). These xenografts that survived long-term express the protective genes heme oxygenase-1 (HO-1), A20, bcl-2, and bcl-x_L in their EC and smooth muscle cells (SMC), whereas xenografts that were rejected hearts did not express these genes or did so weakly (Bach et al. 1997d). We hypothesized that one or more of these protective genes played a functional role in xenograft survival since, when expressed in EC in vitro, these genes prevent EC apoptosis and block the activation of the transcription factor NF-$\varkappa$B, thus preventing the up-regulation of the pro-inflammatory genes associated with EC activation (Bach et al. 1997d).

In mouse hearts transplanted into rats treated with CVF plus CsA, we found that the surviving grafts expressed in their EC and SMC HO-1 but not the other above-mentioned protective genes (Koyamada et al. 1998). As for the other anti-apoptotic genes, HO-1 should be regarded as a protective gene since it blocks EC apoptosis and inhibits the expression of pro-inflammatory genes associated with EC activation. Thus this xenograft model allowed us to test critically whether the expression of a single protective gene in a cardiac xenograft, HO-1, could be shown to be important in graft survival. This could be tested directly, since K. D. Poss and S. Tonegawa had derived a HO-1-deficient mouse (HO-1$^{-/-}$) strain that could be used as a heart donor for transplantation in rats (Poss and Tanegawa 1997a, b; Soares et al. 1998b).

Accommodation of Concordant Xenografts is Dependent on the Expression of the Protective Gene HO-1

Mouse hearts transplanted into untreated rats are rejected three to five days after transplantation. Treatment of the recipient with CsA or CVF alone does not suppress rejection whereas treatment with a combination of CsA plus CVF suppresses rejection and prolongs xenograft survival indefinitely. Xenografts harvested three days after transplantation under CsA plus CVF treatment show normal myocardial morphology associated with only focal and mild infiltration with monocyte/macrophages and no signs of thrombosis. Mouse hearts transplanted into CsA plus CVF-treated rats expressed the protective gene HO-1 in EC and smooth muscle cells as early as 12–24 hours after transplantation.

To test whether the expression of HO-1 was functionally associated with xenograft survival, we transplanted hearts from mice homozygotis (HO-1$^{-/-}$) or heterozygous (HO-1$^{+/-}$) deficient for HO-1 as well as from normal wild type (HO-1$^{+/+}$) mice on the same or another genetic background to rats treated with CVF and CsA. All hearts from wild type HO-1$^{+/+}$ mice accommodated (n = 21) independently of the strain background used. All hearts from homozygous-deficient HO-1$^{-/-}$ hearts (n = 6) were rejected in three to five days. Hearts from heterozygous HO-1$^{+/-}$ accommodated in four of six cases, while the other two were rejected seven and eight days after transplantation.

The pathological features of HO-1$^{-/-}$ hearts rejected in recipients treated with CVF plus CsA are very similar to of HO-1$^{+/+}$ hearts rejected in untreated recipients. There is thrombosis of the coronary vascular tree associated with host monocyte/macrophage infiltration and tissue necrosis. In the absence of HO-1 (HO-1$^{-/-}$ hearts), there are additional pathological features that are not observed in HO-1$^{+/+}$ hearts rejected in untreated recipients. These include thrombosis of

large coronary arteries, myocardial infarction and necrosis and apoptosis of cardiac myocytes. These data suggest that early expression of HO-1 after transplantation may be needed to prevent thrombosis of the coronary vascular tree. When
HO is not present or under inhibition of HO activity, there is widespread coagulation in these vessels and presumably a decrease of blood flow to the heart, leading to anoxia, myocardial infarction, necrosis and apoptosis (Soares et al. 1998b).

   In an attempt to differentiate possible protective effects of HO-1 on non-
immunological factors leading to rejection and a specific immunological effect,
we transplanted HO-1$^{-/-}$ hearts into allogeneic RAG-2-deficient (RAG-2$^{-/-}$) mice.
which lack T and B cells. Both HO-1$^{+/+}$ (n = 2) and HO-1$^{-/-}$ (n = 4) hearts survived long-term in the RAG-2$^{-/-}$ mice, with or without CsA and CVF treatment.
These data suggest, but do not prove, that HO-1 in the mouse-to-rat model inhibits an immune-mediated type of xenograft rejection. Finally these studies provided the first evidence for the expression of a protective gene to assure xenograft survival (Soares et al. 1998b).

An Overview of these Results

The classical reaction of EC during activation is described above. With regard to
gene induction, it is the pro-inflammatory genes that have received attention
since the initial description of EC activation. Our data and those of others now
show that as a part of EC activation, there can be the induction of protective
genes. From in vitro work it is clear that overexpression of these genes can inhibit
the induction of the pro-inflammatory genes. To the extent that one or more protective genes can inhibit the induction of the pro-inflammatory genes in vivo, the
protective gene response would represent a mechanism to prevent an unfettered
pro-inflammatory response.

   The protein products of the pro-inflammatory genes can in some cases
induce EC death-apoptosis. Since the protective genes studied to date can all
inhibit apoptosis as well, their dual action could represent a two-pronged attack
to assure that EC apoptosis will not take place. They would inhibit the generation
of substances as a part of the pro-inflammatory response that could lead to apoptosis and well as acting within the cell to block apoptosis. They truly thus "protect" the EC.

   It is not established, as alluded to above, that the expression of the protective
genes in vivo, which we have shown occurs with xenograft survival, will have the
same functions that have been demonstrated for them in vitro. It seems most
likely, however, that their overexpression in vivo for therapeutic purposes would
suppress the pro-inflammatory response and have the anti-apoptotic effect. It is in
this vein that we are considering their use to aid in achieving xenograft survival.

   We speculate that the protective gene responses, and their use therapeutically, might have more broad-ranging implications than for xenotransplantation
alone. In vascular disease in general, there are times when protective gene function might be therapeutically beneficial. Also, it seems reasonable to at least

speculate that the dual functions of these genes could provide relief in other clinical situations in which certain cells are highly activated and causing damage resulting in disease.

We hypothesize that the up-regulation of protective genes in EC is a part of a regulated response to injury. The balance of expression of pro-inflammatory and protective genes might allow a pro-inflammatory situation to develop but also to control it, so that the response does not lead to rejection or EC death among other unwanted effects. The recognition of the effects of the protective genes may allow for their use in therapy. For xenotransplantation, it is possible to create transgenic pig donor organs. For human vascular diseases, it may be possible to express these genes in selected vessel segments.

Acknowledgements

This is paper #761 from our laboratories. This work was supported by grant HL58688 (FHB) from the NIH, and a grant from Novartis Pharma, Basel, Switzerland. FHB is the Lewis Thomas Professor at Harvard Medical School and a paid consultant to Novartis Pharma.

References

Agarwal A, Kim Y, Matas AJ, Alam J, Nath KA (1996) Gas-generating systems in acute renal allograft rejection in the rat – co-induction of heme oxygenase and nitric oxide synthase. Transplantation 61:93–98

Alexandre GP, Squifflet JP, De Bruyere M, Latinne D, Reding R, Gianello P, Carlier M, Pirson Y (1987) Present experiences in a series of 26 ABO-incompatible living donor renal allografts. Transplant Proc 19:4538–4542

Anrather J, Csizmadia V, Brostjan C, Soares MP, Bach FH, et al. (1997) Inhibition of bovine endothelial cell activation in vitro by regulated expression of a transdominant inhibitor of NF-κB. J Clin Invest 99:763–772

Bach FH, Robson SC, Ferran C, Winkler H, Millan MT, Stuhlmeier KM, Vanhove B, Blakely ML, van der Werf WJ, Hofer E, De Martin R, Hancock WW (1994) Endothelial cell activation and thromboregulation during xenograft rejection. Immunol Rev 141:5–30

Bach FH, Winkler H, Ferran C, Hancock WW, Robson SC (1996) Delayed xenograft rejection. Immunol Today 17:379–384

Bach FH, Hancock WW, Ferran C (1997a) Protective genes expressed in endothelial cells – a regulatory response to injury. Immunol Today 18:483–486

Bach FH, Ferran C, Soares M, Wrighton CJ, Anrather J, Winkler H, Robson SC, Hancock WW (1997b) Modification of vascular responses in xenotransplantation – inflammation and apoptosis. Nature Med 3:944–948

Bach FH, Ferran C, Soares M, Wrighton CJ, Anrather J, Winkler H. Robson SC, Hancock WW (1997c) Modification of vascular responses in xenotransplantation: inflammation and apoptosis. Nature Med 3:944–948

Bach FH, Ferran C, Hechenleitner P, Mark W, Koyamada N, Miyatake T, Winkler H, Badrichani A, Candinas D, Hancock WW (1997d) Accommodation of vascularized xenografts: expression of "protective genes" by donor endothelial cells in a host Th2 cytokine environment. Nature Med 3:196–204

Badrichani AZ, Stroka DM, Bilbao G, Curiel DT, Bach FH, Ferran C (1999) Bcl-2 and Bcl-XL serve an anti-inflammatory function in endothelial cells through inhibition of NF-κB. Journal of Clinical Investigation 103 (4):543–533

Baichwal VR, Baeuerle PA (1997) Activate NF-ϰB or die? Curr Biol 7:R94–96

Balla G, Jacob HS, Balla J, Rosenberg M, Nath K, Apple F, Eaton JW, Vercellotti GM (1992) Ferritin: a cytoprotective antioxidant strategem of endothelium. J Biol Chem 267:18148–18153

Balla J, Jacob HS, Balla G, Nath K, Eaton JW, Vercelotti GM (1993) Endothelial-cell heme uptake from heme proteins: induction of sensitization and desensitization to oxidant damage. Proc Nat Acad Sci USA 90:9285–9289

Bannett AD, McAlack RF, Raja R, Baquero A, Morris M (1987) Experiences with known ABO-mismatched renal transplants. Transplant Proc 19:4543–4546

Beg AA, Baltimore D (1996) An essential role for NF-ϰB in preventing TNF-α-induced cell death. Science 274:782–784

Blakely ML, Van der Werf WJ, Berndt MC, Dalmasso AP, Bach FH, Hancock WW (1994) Activation of intragraft endothelial and mononuclear cells during discordant xenocraft rejection. Transplantation 58:1059–1066

Buttke TM, Sandstrom PA (1994) Oxidative stress as a mediator of apoptosis. Immunol Today 15:7–10

Candinas D, Bach FH, Hancock WW (1996) Delayed xenograft rejection in complement-depleted T-cell-deficient rat recipients of guinea pig cardiac grafts. Transplantation Proceedings 28 (2)

Choi AM, Alam J (1996) Heme oxygenase-1: function, regulation, and implication of a novel stress-inducible protein in oxidant-induced lung injury. Am J Respir Cell Mol Biol 15:9–19

Chu ZL, McKinsey TA, Liu L, Gentry JJ Malim MH et al. (1997) Suppression of tumor necrosis factor-induced cell death by inhibitor of apoptosis c-IAP2 is under NF-kappaB control. Proc Nat Acad Sci USA 94:10057–10062

Collins BH, Cotterell AH, McCurry KR, Alvarado CG, Magee JC, Parker W, Platt JL (1995a) Cardiac xenografts between primate species provide evidence for the importance of the alpha-galactosyl determinant in hyperacute rejection. J Immunol 154:5500–5510

Collins T, Read MA, Neish AS, Whitley MZ, Thanos D, et al. (1995b) Transcriptional regulation of endothelial cell adhesion molecules: NF-ϰB and cytokine-inducible enhancers. FASEB J 9:899–909

Cooper JT, Stroka DM, Brostjan C, Palmetshofer A, Bach FH, Ferran C (1996) A20 blocks endothelial cell activation through a NF-ϰB-dependent mechanism. J Biol Chem 271:18068–18073

Cotran RS, Pober JS (1989) Effects of cytokines on vascular endothelium: their role in vascular and immune injury. Kidney Int 35: 969–975

Eisenstein RS, Garcia MD, Pettingell W, Munro HN (1991) Regulation of ferritin and heme oxygenase synthesis in rat fibroblasts by different forms of iron. Proc Nat Acad Sci USA 88:688–692

Ferran C, Stroka DM, Badrichani AZ, Cooper JT, Wrighton CJ, Soares M, Grey ST, Bach FH (1998) A20 inhibits NF-ϰB activation in endothelial cells without sensitizing to tumor necrosis factor-mediated apoptosis. Blood 91:2249–2258

Galili U (1993) Interaction of the natural anti-Gal antibody with alpha-galactosyl epitopes: a major obstacle for xenotransplantation in humans. Immunol Today 14:480–482

Galili U, Swanson K (1991) Gene sequences suggest inactivation of alpha-1,3-galactosyltransferase in catarrhines after the divergence of apes from monkeys. Proc Natl Acad Sci USA 88:7401–7404

Galili U, Rachmilewitz EA, Peleg A, Flechner I (1984) A unique natural human IgG antibody with anti-alpha-galactosyl specificity. J Exp Med 160:1519–1531

Galili U, Macher BA, Buehler J, Shohet SB (1985) Human natural anti-alpha-galactosyl IgG II. The specific recognition of alpha (1–3)-linked galactose residues. J Exp Med 162:573–582

Galili U, Mandrell RE, Hamadeh RM, Shohet SB, Griffiss JM (1988) Interaction between human natural anti-alpha-galactosyl immunoglobulin G and bacteria of the human flora. Infect Immunol 56:1730–1737

Good AH, Cooper DK, Malcolm AJ, Ippolito RM, Koren E, Neethling FA, Ye Y, Zuhdi N, Lamontagne LR (1992) Identification of carbohydrate structures that bind human antiporcine antibodies: implications for discordant xenografting in humans. Transplant Proc 24:559–562

Hancock WW, Buelow R, Sayegh MH, Turka LA (1998) Antibody-induced transplant arteriosclerosis is prevented by graft expression of anti-oxidant and anti-apoptotic genes. Nature Med 4:1392–1396

Harrison PM, Arosio P (1996) Ferritins – molecular properties, iron storage function and cellular regulation. Biochim Biophys Acta – Bioenergetics 1275:161–203

Hasan R, Van den Bogaerde, JB, Wallwork J, White DJ (1992) Evidence that long-term survival of concordant xenografts is achieved by inhibition of antispecies antibody production. Transplantation 54:408–413

Hattori R, Hamilton KK, McEver RP, Sims PJ (1989a) Complement proteins C5b-9 induce secretion of high molecular weight multimers of endothelial von Willebrand factor and translocation of granule membrane protein GMP-140 to the cell surface. J Biol Chem 264:9053–9060

Hattori R, Hamilton KK, Fugate RD, McEver RP, Sims PJ (1989b) Stimulated secretion of endothelial von Willebrand factor is accompanied by rapid redistribution to the cell surface of the intracellular granule membrane protein GMP-140. J Biol Chem 264:7768–7771

Ishikawa K, Navab M, Leitinger N, Fogelman AM, Lusis AJ (1997) Induction of heme oxygenase-1 inhibits the monocyte transmigration induced by mildly oxidized LDL. J Clin Invest 100:1209–1216

Karsan A, Yee E, Harlan JM (1996) Endothelial cell death induced by tumor necrosis factor-alpha is inhibited b the Bcl-2 family member, A1. J Biol Chem 271:27201–27204

Kobayashi T, Taniguchi S, Neethling FA, Rose AG, Hancock WW, Ye Y, Niekrasz M, Kosanke S, Wright LJ, White DJ, Cooper DK (1997) Delayed xenograft resection of pig-to-baboon cardiac transplants after cobra venom factor therapy. Transplantation 64:1255–1261

Koyamada N, Miyatake T, Candinas D, Mark W, Hechenleitner P, Hancock WW, Soares MP, Bach FH (1998) Transient complement inhibition plus T-cell immunosuppression induces long-term survival of mouse-to-rat cardiac xenografts. Transplantation 65:1210–1215

Lambrigts D, Sachs DH, Cooper DKC (1998) Discordant organ xenotransplantation in primates – world experience and current status. Transplantation 66:547–561

Latinne D, Squifflet JP, De Bruyere M, Pirson Y, Gianello P, Sokal G, Alexandre GP (1989) Subclasses of ABO isoagglutinins in ABO-incompatible kidney transplantation. Transplant Proc 21:641–642

Lee PJ, Alam J, Wiegand GW, Choi AM (1996) Overexpression of heme oxygenase-1 in human pulmonary epithelial cells results in cell growth arrest and increased resistance to hyperoxia. Proc Nat Acad Sci USA 93:10303–10398

Lesnikoski BA, Shaffer DA, Van der Werf WJ, Dalmasso AP, Soares MP, Latinne D, Bazin H, Hancock WW, Bach FH (1995) Endothelial and host mononuclear cell activation and cytokine expression during rejection of pig-to-baboon discordant xenografts. Transplant Proc 27:290–291

Leventhal JR, Sakiyalak P, Witson J, Simone P, Matas AJ, Bolman RM, Dalmasso AP (1994) The synergistic effect of combined antibody and complement depletion on discordant cardiac xenograft survival in nonhuman primates. Transplantation 57:974–978

Lin SS, Kooyman DL, Daniels LJ, Daggett CW, Parker W, Lawson JH, Hoopes CW, Gullotto C, Li L, Birch P, Davis RD, Diamond LE, Logan JS, Platt JL (1997) The role of natural anti-Gal-Alpha-1–3gal antibodies in hyperacute rejection of pig-to-baboon cardiac xenotransplants. Transplant Immunol 5:212–218

Lin SS, Widner BC, Byrne GW, Diamond LE, Lawson JH, Hoopes CW, Daniels LJ, Daggett CW, Parker W, Harland RC, Davis RD, Bollinger RR, Logan JS, Platt JL (1998) The role of antibodies in acute vascular rejection of pig-to-baboon cardia transplants. J Clin Invest 101:1745–1756

Lozada C, Levin RI, Huie M, Hirschhorn R, Naime D, Whitlow M, Recht PA, Golden B, Cronstein BN (1995) Identification of C1q as the heat-labile serum cofactor required for immune complexes to stimulate endothelial expression of the adhesion molecules E-selectin and intercellular and vascular cell adhesion molecules 1. Proc Natl Acad Sci USA 92:8378–8382

Maines MD (1997) The heme oxygenase system: a regulator of second messenger gases. Ann Rev Pharmacol Toxicol 37:517–554

Mantovani A, Bussolino F, Introna M (1997) Cytokine regulation of endothelial cell function – from molecular level to the bedside. Immunol Today 18:231–240

McCurry KR, Parker W, Cotterell AH, Weidner BC, Lin SS, Daniels LJ, Holzknecht ZE, Byrne GW, Diamond LE, Logan JS, Platt JL (1997) Humoral responses to pig-to-baboon cardiac transplantation – implications for the pathogenesis and treatment of acute vascular rejection and for accommodation. Human Immunol 58:91–105

Miyatake T, Sato K, Takigami K, Koyamada N, Hancock WW, Bazin H, Latinne D, Bach FH, Soares MP (1998) Complement-fixing elicited antibodies are a major component in the pathogenesis of xenograft rejection. Journal of Immunology 160 (8):4114–23

Morita T, Kourembanas S (1995) Endothelial cell expression of vasoconstrictors and growth factors is regulated by smooth muscle cell-derived carbon monoxide. J Clin Invest 96:2676–2682

Opipari AJ, Hu HM, Yabkowitz R, Dixit VM (1992) The A20 zinc finger protein protects cells from tumor necrosis factor cytotoxicity. J Biol Chem 267:12424–12427

Otterbein LE, Bach FH, Alam J, Soares M, Tao HL, Wysk M, Davis R, Flavell R, A.M.K. C (2000) Carbon monoxide mediates anti-inflammatory effects via the mitogen activated protein kinase pathway. Nature Medicine in press

Otterbein L, Sylvester SL, Choi AM (1995) Hemoglobin provides protection against lethal endotoxemia in rats: the role of heme oxygenase-1. Am J Respir Cell Mol Biol 13:595–601

Otterbein L, Chin BY, Otterbein SL, Lowe VC, Fessler HE, Choi AM (1997) Mechanism of hemoglobin-induced protection against endotoxemia in rats: a ferritin-independent pathway. Am J Physiol 272:L268–275

Parker W, Saadi S, Lin SS, Holzknecht ZE, Bustos M, Platt JL (1996) Transplantation of discordant xenografts: a challenge revisited. Immunol Today 17:373–378

Petrache I, Alam J, Choi AMK (1997) Heme oxygenase 1 (HO-1) inhibits tumor necrosis factor-α (TNF-α)-induced apoptosis in L929 cells. In: Cold Spring Harbor Seminar: Apoptosis. Cold Spring Habor, New York

Picard V, Epszteijn S, Santambrogio P, Cabantchik ZI, Beaumont C (1998) Role of Ferritin in the control of the labile iron pool in murine erythroleukemia cells. J Biol Chem 273:15382–15386

Platt JL (1998) New directions for organ transplantation. Nature 392:11–17

Platt JL, Nath KA (1998) Heme oxygenase: protective gene or Trojan horse. Nature Med 4:1364–1365

Platt JL, Vercellotti GM, Lindman BJ, Oegema TR, Jr, Bach FH, Dalmasso AP (1990) Release of heparan sulfate from endothelial cells. Implications for pathogenesis of hyperacute rejection. J Exper Med 171:1363–1368

Pober JS, Cotran RS (1990) The role of endothelial cells in inflammation. Transplantation 50:537–544

Poss KD, Tonegawa S (1997a) Heme oxygenase 1 is required for mammalian iron reutilization. Proc Nat Acad Sci USA 94:10919–10924

Poss KD, Tonegawa S (1997b) Reduced stress defense in heme oxygenase 1-deficient cells. Proc Nat Acad Sci USA 94:10925–10930

Pruitt SK, Bollinger RR, Collins BH, Marsh HJ, Levin JL, Rudolf AR, Baldwin WW, Sanfilippo F (1997) Effect of continuous complement inhibition using soluble complement receptor type 1 on survival of pig-to-primate cardiac xenografts. Transplantation 63:900–902

Reinders JH, de Groot PG, Sixma JJ, van Mourik JA (1988) Storage and secretion of von Willebrand factor by endothelial cells. Haemostasis 18:246–261

Robson SC, Kaczmarek E, Siegel JB, Candinas D, Koziak K, Millan M, Hancock WW, Bach FH (1997) Loss of ATP diphosphohydrolase activity with endothelial cell activation. J Exper Med 185:153–163

Roy N, Deveraux QL, Takahashi R, Salvesen GS, Reed JC (1997) The c-IAP-1 and c-IAP-2 proteins are direct inhibitors of specific caspases. EMBO J 16:6914–6925

Saadi S, Platt JL (1995) Transient perturbation of endothelial integrity induced by natural antibodies and complement. J Exper Med 181:21–31

Saadi S, Holzknecht RA, Patte CP, Stern DM, Platt JL (1995) Complement mediated regulation of tissue factor activity in endothelium. J Exp Med 182:1807–1814

Scarpati EM, Sadler JE (1989) Regulation of endothelial cell coagulant properties. Modulation of tissue factor, plasminogen activator inhibitors, and thrombomodulin by phorbol 12-myristate 13-acetate and tumor necrosis factor. J Biol Chem 264:20705–20713

Slapak M, Naik RB, Lee HA (1981) Renal transplant in a patient with major donor-recipient blood group incompatibility: reversal of acute rejection by the use of modified plasmapheresis. Transplantation 31:4–7

Soares MP, Brouard S, Smith RN, Otterbein L, Choi AM, Bach FH (2000) Expression of heme oxygenase-1 by endothelial cells: a protective response to injury in transplantation. Emerging therapeutic targets 4 (1):11–27

Soares MP, Lin Y, Sato K, Stuhlmeier KM, Bach FH (1999) Accommodation. Immunology Today 20 (10):434–437 (1999 b)

Soares MP, Lin Y, Sato K, Takigami K, Anrather J, Ferran C, Robson SC, Bach FH (1999) Pathogenesis of and potential therapies for delayed xenograft rejection. Current opinion in organ transplantation 4:80–89

Soares MP, Muniappan A, Kaczmarek E, Koziak K, Wrighton CJ, et al. (1998a) Adenovirus mediated expression of a dominant negative mutant of p65/RelA inhibits proinflammatory gene expression in endothelial cells without sensitizing to apoptosis. J Immunol 161:4572–4582

Soares MP, Lin Y, Anrather J, Csizmadia E, Takigami K, Sato K, Grey ST, Colvin RB, Choi AM, Poss KD, Bach FH (1998b) Expression of heme oxygenase-1 (HO-1) can determine cardiac xenograft survival. Nature Med 4:1073–1077

Stocker R, Yamamoto Y, McDonagh AF, Glazer AN, Ames BN (1987) Bilirubin is an antioxidant of possible physiological importance. Science 235:1043–1046

Stroka DM, Badrichani AZ, Bach FH, Ferran C (1999) Overexpression of A1, an NF-κB-inducible anti-apoptotic bcl gene, inhibits endothelial cell activation. Blood 93 (11):3803–3810

Tedesco F, Pausa M, Nardon E, Introna M, Mantovani A, Dobrina A (1997) The cytolytically inactive terminal complement complex activates endothelial cells to express adhesion molecules and tissue factor procoagulant activity. J Exper Med 185:1619–1627

Utgaard JO, Jahnsen FL, Bakka A, Brandtzaeg P, Haraldsen G (1998) Rapid secretion of prestored interleukin 8 from Weibel-Palade bodies of microvascular endothelial cells. J Exper Med 188:1751–1756

Vanantwerp DJ, Martin SJ, Kafri T, Green DR, Verma IM (1996) Suppression of TNF-α-induced apoptosis by NF-κB. Science 274:787–789

van den Berg RH, Faber-Krol MC, Sim RB, Daha MR (1998) The first subcomponent of complement, C1q, triggers the production of IL-8, IL-6, and monocyte chemoattractant peptide-1 by human umbilical vein endothelial cells. J Immunol 161:6924–6930

Vanhove B, Bach FH (1993) Human xenoreactive natural antibodies – avidity and targets on porcine endothelial cells. Transplantation 56:1251–1253

Wagner CT, Durante W, Christodoulides N, Hellums JD, Schafer AI (1997) Hemodynamic forces induce the expression of heme oxygenase in cultured vascular smooth muscle cells. J Clin Invest 100:589–596

Wang CY, Mayo MW, Baldwin AS (1996) TNF- and cancer therapy-induced apoptosis – potentiation by inhibition of NF-κB. Science 274:784–787

Willis D, Moore AR, Frederick R, Willoughby DA (1996) Heme oxygenase: a novel target for the modulation of the inflammatory response. Nature Med 2:87–90

Wolff B, Burns AR, Middleton J, Rot A (1998) Endothelial cell memory of inflammatory stimulation – human venular endothelia cells store interleukin 8 in Weibel-Palade bodies. J Exper Med 188:1757–1762

Wong GH, Elwell JH, Oberley LW, Goeddel DV (1989) Manganous superoxide dismutase is essential for cellular resistance to cytotoxity of tumor necrosis factor. Cell 58:923–931

Wrighton CJ, Hofer WR, Moll T, Eytner R, Bach FH, de Martin R (1996) Inhibition of endothelial cell activation by adenovirus-mediated expression of IκBα, an inhibitor of the transcription factor NF-κB. J Exp Med 183:1013–1022

Wu MX, Ao Z, Prasad KV, Wu R, Schlossman SF (1998) IEX-1L, an apoptosis inhibitor involved in NF-κB-mediated cell survival. Science 281:998–1001

Xu H, Gundry SR, Hancock WW, Matsumiya G, Zuppan CW, Morimoto T, Slater J, Bailey LL (1998) Prolonged discordant xenograft survival and delayed xenograft rejection in a pig-to-baboon orthotopic cariac xenograft model. J Thoracic Cardiovasc Surg 115:1342–1349

Yachie A, Niida Y, Wada T, Igarashi N, Kaneda H, Toma T, Ohta K, Kasahara Y, Koizumi S (1999) Oxidative stress causes enhanced endothelial cell injury in human heme oxygenase-1 deficiency. J Clin Invest 103:129–135

Genetic Factors in Malaria Resistance

L. Luzzatto

A classical representation of the causation of human diseases is one in which every disease has both an inherited and an acquired component; it is only the relative weights of the two components that vary. Infectious diseases have lent themselves well to illustrating this notion, as their etiology is unquestionably an exogenous agent, but the pathogenesis is more or less strongly influenced by the biology of the host, which is in turn to a large extent genetically determined. The interaction between the host and the infectious agent operates in two ways: on one band, the infectious agent may encounter pre-existing host defenses (e.g., the immune system); on the other hand, by being unable to kill resistant hosts, it favors the propagation of inheritable resistance factors.

Malaria May Have Been a Major Selective Force in Human Evolution

Among infectious diseases, malaria is one of the best examples of how a parasite can exert a strong selective pressure on its human host, for at least four reasons: 1) *Plasmodium falciparum* infection is highly lethal, and most of the deaths occur in the pre-reproductive age (Bradley et al. 1996); 2) it is estimated that *P falciparum* malaria has become endemic or hyper-endemic in large tropical and sub-tropical parts of the world with the introduction of agriculture, which produced a marked increase in the density of human populations (O'Donnell et al. 1998). Therefore, unlike with transient epidemics, there has been an opportunity for continuous selection of inherited resistance factors over a period of up to several hundred generations. Today, *P falciparum* malaria is still rampant in many tropical and sub-tropical parts of the world (see Fig. 1). It is estimated that about 500 million people still live in malaria-endemic areas and that malaria is responsible for 1–3 million deaths/year; 3) both the life cycle (see Fig. 2) and the mechanism whereby the malaria parasite causes death are complex. Therefore we can visualize potential mechanisms of resistance (Luzzatto 1979) at the level of several distinct targets within the host, including the following: the liver cell, the erythrocyte, the immune system, as well as local or humoral factors that may influence the development of cerebral malaria, a major cause of mortalitys; 4) among these targets, the red cells are exquisitely amenable to investigation. Therefore, an unusually large body of detailed information bas accumulated on their genetics, biochemistry and physiology (Luzzatto 1985).

V. Boulyjenkov, K. Berg, Y. Christen (Eds.)
Genes and Resistance to Diseases
© Springer-Verlag Berlin Heidelberg 2000

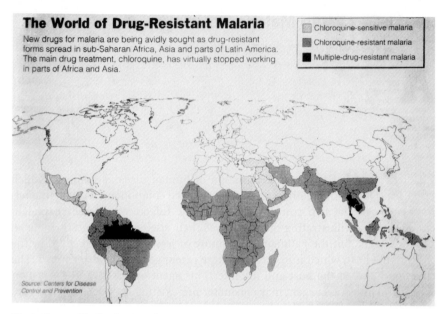

Fig. 1. Geographic distribution of malaria, highlighting the major problem of drug resistance, mainly important with respect to *P falciparum*. Until the mid-1960s malaria responded practically always to chloroquine, which was thus a very effective and very safe curative drug for this condition. Now the rate of chloroquine resistance in the dark-shaded areas ranges from about 10 to 90 %. This map was published in The New York Times in 1995.

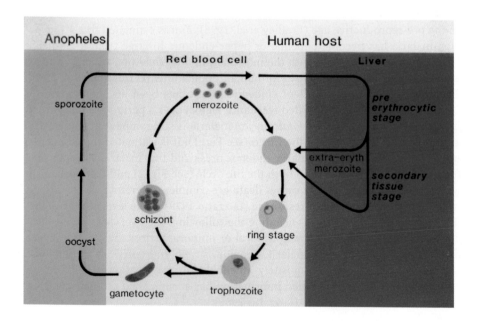

What Approaches Can Be Used to Investigate Genetic Resistance to Malaria?

Several speakers in this symposium have observed that traditionally there has been greater emphasis on identifying genetic factors that increase susceptibility to a disease, rather than factors that increase resistance. By contrast, *P falciparum* is such a dangerous parasite that one might be tempted to say that dying from this form of malaria is the rule rather than the exception. As a result, the search for genetic resistance factors has been active since JBS Haldane put forward this notion (Haldane 1949). A variety of different approaches have been used to show that people with a particular genotype have increased resistance to *P falciparum* malaria (Table 1).

Table 1. Possible mechanisms of protection against malarial Plasmodia in red blood cells

At the invasion stage	*Failure of infection*
During intra-erythrocytic development	*Abortive infection*
Removal of parasitized cells	*Suicidal infection*

What Genes May Have Been Subject to Selection?

There is now compelling evidence that genetic resistance factors have been selected by *P falciparum* within all three major components of the organization of the red cells: 1) hemoglobin, 2) metabolic pathways, and 3) the membrane-cytoskeleton complex.

1) Both structural and quantitative abnormalities in the hemoglobin (Hb) System can affect resistance to *P falciparum* Hb S is of course the classical example of the former (it is more doubtful that Hb C protects from malaria); both α-

---

Fig. 2. The life cycle of the malaria parasite. In the case of *P falciparum* the sporozoites injected by the mosquito disappear from the blood stream into the liver within minutes, thanks to the high affinity of the thrombospondin-related anonymous protein (TRAP) on their surface for the surface of liver cells (Sultan et al. 1997). The parasites then multiply within liver cells, and it is estimated that each sporozoite can yield up to 30,000 pre-erythrocytic merozoites, ready to invade red cells. From then on the main goings-on are in or related to red cells. At each cycle the multiplication factor is up to 32, and this causes rapidly increasing destruction of red cells, with consequent anemia (Luzzatto 1979) (although there are several other factors that contribute to the anemia: see Weatherall et al. 1983). The surface of the red cells is modified, which may affect their ability to move through capillaries, adhesion to endothelial cells, etc. These phenomena probably play a role in the pathogenesis of cerebral malaria (Bradley et al. 1996; Newton and Warrell 1998). The parasitized red cells can be recognized by macrophages as abnormal and therefore removed, with destruction of the parasite: the stage at and the efficiency with which this happens are critical to the outcome of the infection. Thus, most of the pathology associated with *P falciparum* malaria, and specifically the two most fatal complications (severe anemia and cerebral malaria), depend on the parasite's intra-erythrocytic forms. Therefore, it stands to reason that genetic changes expressed in red cells, even if they have only a small effect on each cycle, may ultimately have a major effect on the outcome of a malaria attack. At each cycle the parasite has the option of sexual differentiation, thus interrupting the highly pathogenic cycle but producing gametocytes capable of re-infecting the mosquito and thus maintaining malaria transmission in the community.

and β-thalassemia are examples of the latter (some structurally abnormal hemo-globins such as Hb E and Hb Lepore are also protective, probably because they are associated with a thalassemia phenotype). 2) Glucose 6-phosphate dehydro-genase (G6PD) is the first and rate-limiting enzyme of the pentose phosphate pathway: in the red cell, by virtue of its ability to generate NADPH, it is the main defense against oxidative stress. Numerous G6PD deficiency mutants have become prevalent in many human populations as a result of malaria selection. 3) Among membrane proteins, structural abnormalities of the anion channel called band 3 cause changes in membrane rigidity and in the overall shape of the red cells (ovalocytosis; Holt et al. 1981). Ovalocytosis has been probably selected by *P. falciparum* (Liu et al. 1995; Schofield et al. 1992; Nurse et al. 1992). The Duffy-null gene *Fy* impedes penetration of *P. vivax* (Maugh, 1981), but it is doubtful that it confers resistance against *P. falciparum*. Since *P. vivax* is not endemic wherever *Fy* is prevalent, I rather favor the view that in this case the high preva-lence of this gene has prevented the spread of this parasite, rather than the reverse.

What Mechanisms May Be Involved in Genetic Resistance to Malaria?

First of all, it is important to state that "resistance" to malaria is probably always relative rather than absolute. We do not know of any single factor, other than lack of exposure, that can prevent a person from developing malaria. Therefore, we use the phrase (relative) resistance to indicate that a person has a decreased risk of suffering from the more severe clinical manifestations of malaria, and conse-quently a decreased risk of dying from it. Given the complexity of the parasite cycle, several different mechanisms of resistance can be envisaged, both at the level of the red cell (see Table 2) and elsewhere (see Table 1). Acquired immunity against malaria is highly complex, as shown by the fact that development of acquired resistance requires multiple attacks over a period of years, the immu-nity generated is relative rather than absolute, and it can break down in certain situations (e.g., pregnancy, lack of exposure). Although very numerous anti-bodies are produced, it appears that cellular immunity may be more important for protection. At least one HLA allele has been selected by *P falciparum*. Recent evidence suggests that non-MHC-restricted phospholipid antigens may be important in immunity against parasitic protozoa (Schofield et al. 1999). The rel-

Table 2. Red cell traits that confer resistance to malaria

	Evidence obtained in			
	AS	α-thal	β-thal	G6PD deficiency
Epidemiology	+	+	+	+
Micro-mapping	+	+	+	+
Clinical/field work	+	+		+
In vitro cultures	+	(+)		+
Cellular/molecular mechanism	+	(+)		+

evant surface molecule is CD1 (Joyce et al. 1998), which is not known to be poly-morphic: it might be interesting to investigate possible genetic variation of the respective genes in populations living in malaria-hyperendemic areas.

Thus, a variety of diverse genes may offer some protection against malaria, a variety of diverse approaches have been used to investigate this clinically impor-tant phenomenon, and the underlying biological mechanisms are also quite diverse. Time does not allow us to cover the evidence that pertains to each case. Therefore, in this review I shall confine myself to some specific examples: natu-rally, my choice of examples is influenced by work that I have been directly involved with over the years.

α-Thalassemia

The Terai region of Nepal, located to the South of the foothills of the Himalayas (see Fig. 3), has been known to be heavily infested by malaria since remote times. Therefore, it has been regarded as virtually uninhabitable by most Nepalese peo-ple. As the only exception, the Tharu people have been living in the Terai for cen-turies, and they were reputed to have an innate resistance to malaria. Since about 1950 the Nepal Malaria Eradication Organization (NMEO) has embarked on a major effort to eradicate malaria, which has achieved a large measure of success. Largely as a result of this effort, a large and heterogeneous non-Tharu population now inhabits the Terai along with Tharus. This rather unique demographic situa-

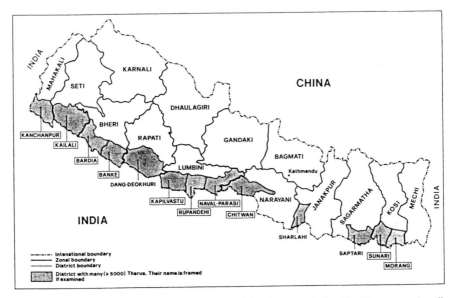

Fig. 3. Map of the Terai region of Nepal, where most of the Tharu people live. The Tharu were virtually the only inhabitants of the region, probably for centuries, before a malaria control campaign was initi-ated about 50 years ago.(From Terrenato et al. 1988).

Table 3. Frequencies of α-globin genes in the Terai (from Modiano et al. 1991)

People and Area in Terai (n)	Estimated Haplotype Frequency (mean ± SE)			
	αα	ααα	α-^{I}	α-^{II}
Tharu				
Western (18)	0.28 ± 0.08		0.67 ± 0.08	0.05 ± 0.04
Central (18)	0.17 ± 0.06		0.83	
Non-Tharu				
Western (17)	0.97		0.03 ± 0.03	
Central (17)	0.85 ± 0.06	0.06 ± 0.04	0.09 ± 0.5	

tion made it possible to investigate directly, within the same villages, whether the perceived resistance to malaria of the Tharu people could be objectively confirmed, and to investigate its possible genetic basis. By analyzing NMEO records, we found that the prevalence of cases of residual malaria is nearly seven times lower among Tharus than in sympatric non-Tharus. This difference applied to both *P vivax*, which is now much more common, and *P falciparum* (Terrenato et al. 1988).

In vitro cultures of *P falciparum* in red cells from Tharus versus those from non-Tharu controls failed to reveal any significant difference in either the rate of invasion or the rate of parasite multiplication. When the prevalence of candidate protective genes was measured, there were relatively minor differences between Tharus and non-Tharus in the frequencies of β-thal, β^{S}, G6PD(–), and Duffy (a–b–) in different parts of the Terai region. By contrast, the frequency of α-thalassemia is uniformly high in Tharus, with the majority of them having the homozygous α-/α-genotype (see Table 3) and an overall α-thal gene frequency of 0.8 (Modiano et al. 1991). Thus, it seems clear that in the Terai holoendemic malaria has caused preferential survival of subjects with α-thal, and that this genetic factor has enabled the Tharus as a population to survive for centuries in a malaria-holoendemic area. It can be estimated that the α-thal homozygous state decreases morbidity from malaria by about 10-fold. This is an example of selective evolution toward fixation of an otherwise abnormal gene.

Hemoglobin S

Hemoglobin S has become perhaps the most popular textbook example of balanced polymorphism. The pioneering work of AC Allison in East Africa first suggested a reduction in the severity of malaria in AS heterozygotes compared to normal (AA) subjects. This finding was followed by experiments in which AS heterozygotes were deliberately exposed to bites by *P falciparum*-infected mosquitoes,[1]

---

[1] Although the ethics of such experiments ought to have been questioned, the subjects were closely supervised and treated with chloroquine before any complications set in. At that time chloroquine was 100 % effective in controlling malaria.

and were shown to develop lower levels of parasitemia than control AA subjects (Allison 1954). Subsequently, several well-controlled clinical studies as well as population-based studies confirmed these findings and showed specifically that the AS heterozygous state does not protect from being infected by *P falciparum* but rather protects from developing those high level of parasitemia that are associated with its two major life-threatening complications, namely severe anemia and cerebral malaria (Bradley et al. 1996). The possible mechanism of this protective effect was first investigated by incubating in vitro red cells from naturally infected AS patients in a low-oxygen environment. This study clearly demonstrated an accelerated sickling of parasitized red cells when compared to non-parasitized red cells (Luzzatto et al. 1970). When W Trager revolutionized malaria research by introducing a technique for the continuous in vitro culture of *P falciparum* (Trager and Jensen 1976), these results were confirmed (Roth et al. 1978). Moreover, it was possible to ask in which way sickling might be protective: whether because it inflicted direct damage to the parasite, or because the parasitized sickled cells would become prey of macrophages. In vitro findings (Luzzatto and Pinching 1990) favor the latter view (see Fig. 4), and they are in keeping with the clinical observation that impaired phagocytosis in vivo, for instance in patients who have a functional or anatomical asplenia, is associated with a very high mortality from malaria (Adeloye et al. 1972). Recently the development of mouse models of sickle cell anemia by transgene technology has made it possible to investigate experimentally the Hb S gene as a resistance factor for murine malaria (*P chabaudi* and *P berghei*). Again the removal of parasitized cells by the spleen has emerged as a major mechanism of resistance (Shear et al. 1993).

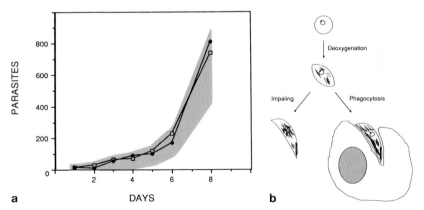

Fig. 4. Mechanism whereby Hb S protects from fatal *P falciparum* malaria. A. The in vitro growth of *P falciparum* is no different in the red cells from two homozygous SS patients (solid lines) than from that observed in red cells from a panel of normal controls (shaded area). B. Cartoon illustrating the fate of parasitized red cells containing Hb S. As the parasite develops it causes a decrease in intracellular p_{O_2} and intracellular pH, favoring Hb S polymerization. This does not seem to cause damage to the parasite directly (in view of the data in panel A); rather it has been shown experimentally that AS parasitized red cells are preferentially phagocytozed by peripheral blood monocytes (and presumably by other macrophages). (Both panels are from Luzzatto and Pinching 1990).

While these data all concur in indicating that Hb AS heterozygotes are at a distinct biological advantage in any area where *P falciparum* is a threat, they do not provide a quantitative estimate of this advantage. To measure the increase in biological fitness conferred upon heterozygotes by the Hb S gene over many past generations, one would have to assess what proportion – without treatment – will die of malaria in an endemic area, compared to controls. Clearly this is not feasible. On the other hand, it is intuitive that the increase in fitness must be substantial, since it must balance the severe decrease in fitness of SS homozygotes, who suffer from sickle cell anemia, and most of whom, without treatment, die before reproduction. On the assumption that the Hb S gene frequency is at equilibrium in populations that have lived for a long time in malaria hyperendemic areas, it has been estimated from population genetics principles that the biological (reproductive) fitness of AS heterozygotes is 20–30 % higher than that of AA subjects (Montanaro et al. 1991). One could say that, in such areas, in terms of health the AS genotype is more "normal" than the AA genotype, and that the Hb S gene has contributed in a major way to the survival of populations in a hazardous environment. There is probably no better example, in the human species, of what is meant by the phrase "genetic load." This is carried entirely by the SS homozygotes, pressing home the responsibility of the community towards them.

G6PD Deficiency

G6PD deficiency is the most widely prevalent enzyme disorder of the red cells (Fig. 5), with population frequencies of 5–20 % in many parts of the world and a peak of over 60 % in an ancestral population of Kurdish Jews (near fixation or founder effect; Oppenheim et al. 1993). The geographic distribution of the G6PD deficient phenotype is remarkably similar to the epidemiology of malaria as it is now or as it was in the recent past (compare Fig. 1 with Fig. 5). Since G6PD-deficient subjects are at risk of severe neonatal jaundice and of hemolytic anemia (Luzzatto and Mehta 1995), it is difficult to imagine that this genetic trait would reach high frequencies unless it also entails some selective advantage. Since 1960 it has been suggested that this advantage might be increased resistance to *P falciparum* malaria (Allison 1960; Motulsky 1960). We now know that different polymorphic mutations underlie the G6PD-deficient phenotype in different populations (Vulliamy et al. 1997). Since these mutations must have arisen independently in genetically disparate people, we can assume that they have also been selected for independently. This assumption further supports the notion that malaria was the selective factor; we have here a good example of evolutionary convergence. Even within a population that is genetically homogeneous, wide variations are observed in the prevalence of G6PD deficiency, depending on the intensity of malaria selection (see Fig. 6; Siniscalco et al. 1966).

Studies in the field have indeed shown that G6PD-deficient children have lower parasite loads (Bienzle et al. 1972, 1979) and a lower incidence of severe malaria (Ruwende et al. 1995) than appropriate controls. In vitro culture work

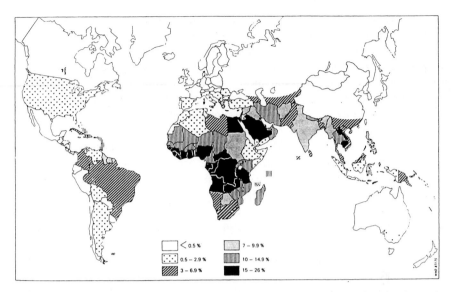

Fig. 5. Epidemiology of G6PD deficiency (from Luzzatto and Mehta 1995). Different polymorphic mutants underlie this phenotype in different parts of the world.

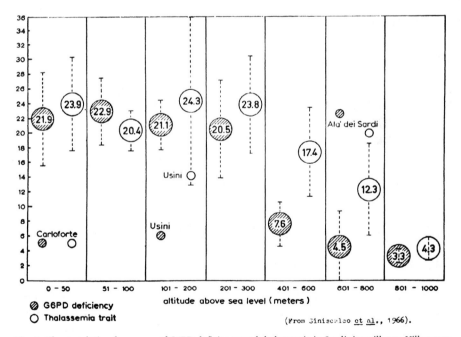

Fig. 6. The population frequency of G6PD deficiency and thalassemia in Sardinian villages. Villages on or near the coast were heavily endemic until malaria eradication was successfully completed in 1948, whereas villages on the highlands were almost free of *P falciparum* malaria. Apart from altitude, these villages are very close and their population is ancestrally the same. This is one of the best examples of micro-mapping as evidence for malaria selection (see Table 1). (From Siniscalco et al. 1966)

has demonstrated impaired growth of *P falciparum* in some studies (Roth et al. 1983; Usanga and Luzzatto 1985) but not in others (Cappadoro et al. 1998). The G6PD gene maps to the X chromosome, and there is some controversy as to whether the relative resistance to this organism is a prerogative of heterozygous females (who are genetic mosaics as a result of X chromosome inactivation; Luzzatto and Mehta 1995) or whether it applies also to hemizygous males (Ruwende et al. 1995). Recently it has been shown that parasitized G6PD-deficient red cells undergo phagocytosis by macrophages at an earlier stage than parasitized normal red cells (Cappadoro et al. 1998), supporting the notion that, in this case, as in AS heterozygotes, suicidal infection is one protective mechanism. Unlike for hemoglobin S, for which there is no counterpart in the parasite, *P falciparum* has its own G6PD gene. While this has significant homology to the host cell gene (see Fig. 7), it also has an additional domain of as yet unknown function (O'Brien et al. 1994). It is possible that, in the context of the intimate relationship between the host red cell and its intracellular parasite, the interplay between the level of G6PD in the former and in the latter also plays a role in malaria resistance (Usanga and Luzzatto 1985; Kurdi-Haidar and Luzzatto 1990).

An intriguing twist to this story is that the very discovery of G6PD deficiency was linked to malaria because the 4-aminoquinoline based antimalarials (such as primaquine) are among the agents that can cause acute hemolytic anemia in subjects who are G6PD deficient (Beutler 1959, 1991). Another trigger of hemolysis is the glycosides present in fava beans (*Vicia faba*), and because this plant is pop-

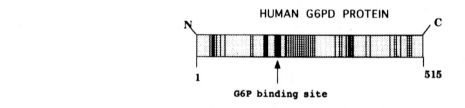

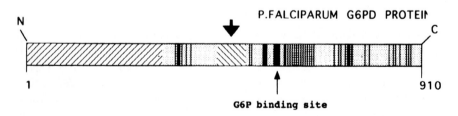

Fig. 7. Sequence comparison of human G6PD and *P falciparum* G6PD. Vertical solid black bars represent regions of high homology/identity; stippled areas indicate regions of lower homology. The hatched N-terminal region and the hatched thick region marked by the thick arrow in *P falciparum*G6PD have no, homology to human G6PD and may fulfill some other function in the parasite protein. (Data from O'Brien et al. 1994)

ular in certain areas, that also have had malaria and G6PD deficiency, a three-way interaction has been suggested (Jackson 1997). In fact it is quite clear that fava beans are not required for selection of G6PD-deficient genes (e.g., they have not been grown traditionally in Africa, where both *P falciparum* and G6PD deficiency are highly endemic). This is not to say that there is no link between all these facts. Indeed, we know that both primaquine and fava beans cause hemolysis because they generate oxygen radicals, and G6PD-deficient red cells are vulnerable to oxidative stress (Arese and De Flora 1990). Although the chemotherapeutic mechanism of action of primaquine against malaria parasites is not full elucidated, oxidative stress may play a role in it (Golenser and Chevion 1989). Finally, it is quite clear that the G6PD-deficient red cell is a sub-optimal host for *P falciparum*, whether because its growth is inhibited or because its removal is accelerated, and this again must be caused by the reduced ability of the G6PD-deficient red cell to withstand oxidative stress.

Conclusion

Most of the information pertaining to these genetic resistance factors has been derived from population genetics and from clinical studies. However, it is gratifying that we are also beginning to have some evidence of the mechanism whereby these genes operate (see Table 4). For instance, in the ease of ovalocytosis, it seems that the changes in the red cell membrane interfere with penetration of the parasite. In the case of Hb S and of G6PD deficiency, the main mechanism of protection seems to be preferential phagocytosis of parasitized red cells. In the case of α-thalassemia, it has been recently suggested that the mechanism of resistance may be, paradoxically, an increased proneness to malaria attacks, possibly favoring the development of immunity. A novel experimental approach, which has now become available for these studies, is the use of mice made transgenic for the relevant human mutant genes (see Shear et al. 1993).

The fact that a variety of human genetic loci affect resistance to malaria bears witness to the impressive power of *P falciparum* as an agent of selection. While this parasite has been perhaps unjustly credited with contributing to the fall of the Roman Empire, it may not be an exaggeration to say that it has played a significant role in shaping human evolution. In this respect, it is important to note that genetic variation is quite extensive not only in *Homo sapiens* but in *P falciparum* as well. Genetic resistance factors in the host may be specific for certain genetic variants of the parasite, and there begins to be evidence that this is indeed the case (Ntoumi et al. 1997). Thus, we may have here a good example not just of a parasite influencing evolution of the host but of the two species actually co-evolving.

Table 4. Mechanism of resistance against *P falciparum* conferred by individual genes

Genetic loci	Mutant allele(s)	Cell type where expressed	Possible mechanism of resistance	Comments	References[a]
β-globin	βˢ	Red cell	**Suicidal infection** (accelerated sickling of parasitized cells facilitates phagocytosis)	Heterozygote advantage balances near-lethality in homozygous βˢ/βˢ	Luzzatto et al. 1970; Friedman 1978; Roth et al. 1978; Luzzatto and Pinching 1990; Shear et al. 1993
	β-thlalassemia	Red cell	Not known (perhaps suicidal)	Many different alleles involved, including β+, β0, HbE, Lepore	Bunyaravej et al. 1986; Luzzi et al. 1991
α-globin	-thalassemia (α-)	Red cell	Paradoxically increased rate of infection improves immune response? Or could be **suicidal**	Mutant allele near fixation in some populations	Luzzi et al. 1991; Williams et al. 1996
Band 3	Ovalocytosis	Red cell	**Reduced invasion**	Polymorphic frequency in SE Asia, Oceania	Kidson et al. 1981; Hadley et al. 1983; O'Donnell et al. 1998
Glucose 6-phosphate dehydrogenase (G6PD)	G6PD deficiency	All cells (housekeeping gene)	**Abortive infection** (growth/developmental arrest after invasion of G6PD-deficient cells); and **suicidal infection** (phagocytosis of parasitized red cells at an earlier stage)	Many alleles have polymorphic frequencies (e.g. G6PD Med in the Mediterranean, G6PD A⁻ in Africa, G6PD Canton in China, etc.)	Luzzatto et al. 1969; Luzzatto 1981; Usanga and Luzzatto 1985; Roth et al. 1983; Cappadoro et al. 1998
MHC Class I	HLA-B53	Most cells	Improved presentation of certain Plasmodial peptides to CD8 T cells?	Protects from "severe" malaria	Hill et al. 1992; Plebanski et al. 1997
MHC class II	HLA-DRB1*1302	Antigen presenting cells	More effective T cell help for certain Plasmodial epitopes?	Different DR alleles may be selected in relation to antigenic variation in parasite	Ohta et al. 1997; Hill 1998
CR1	Sl(a-)	Red cells and other cells	Decreased rosetting through less binding of PfEMP1	Sl(a-) common in Africa but not proven to protect	Rowe et al. 1997
TNF (MHC class III)	-308 promoter mutant[b]	Activated cytotoxic T cells	Increased production of TNF	Homozygous mutant over-represented in cerebral malaria	McGuire et al. 1999
ICAM-1	ICAM-1^Kilifi (K29m)[b]	Endothelial	Unknown	ICAM-1^Kilifi over-represented in cerebral malaria	Fernandez-Reyes et al. 1997

a Mainly on mechanism
b Increased susceptibility rather than increased resistance

Acknowledgements

I am very grateful to all my colleagues with whom I have collaborated in experimental work and in field work in Ibadan, in Napoli, in London and in New York: most of their names are in the reference list. I wish to pay special tribute to Herbert Gilles, who initiated me in malaria research, and to Lucio and Irene Benedetti, who have educated me on the ultrastructure of the red cell.

References

Adeloye A, Luzzatto L, Edington GM (1972) Severe malarial infection in a patient with sickle-cell anaemia. Br Med J 2:445–446

Allison AC (1954) Protection afforded by the sickle cell trait against subtertian malarial infection Br Med J i:290–294

Allison AC (1960) Glucose 6-phosphate dehydrogenase deficiency in red blood cells of East Africans. Nature 186:531–532

Arese P, De Flora A (1990) Pathophysiology of hemolysis in glucose 6-phosphate dehydrogenase deficiency. Sem Hematol 27:1–40

Beutler E (1959) The hemolytic effect of primaquine and related compounds. Blood 14:103–139

Beutler E (1991) Glucose 6-phosphate dehydrogenase deficiency. N Engl J Med 324:169–174

Bienzle U, Ayeni O, Lucas AO, Luzzatto L (1972) Glucose-6-phosphate dehydrogenase deficiency and malaria. Greater resistance of females heterozygous for enzyme deficiency and of males with non-deficient Variant. Lancet i:107–110

Bienzle U, Guggenmoos-Holzmann I, Luzzatto L (1979) Malaria and erythrocyte glucose-6-phosphate dehydrogenase variants in West Africa. Am J Trop Med Hyg 28:619–621

Bradley D, Newbold CI, Warrel DA (1996) Malaria. In: Weatherall DJ, Ledingham JGG, Warrell DA (eds) Oxford textbook of medicine. 3rd ed. Oxford, OUP, pp 835–863

Bunyaratvej A, Butthep P, Yuthavong Y, Fucharoen S, Khusmith S, Yoksan S, Wasi P (1986) Increased phagocytosis of Plasmodium falciparum-infected erythrocytes with haemoglobin E by peripheral blood monocytes. Acta Haematol 76:155–158

Cappadoro M, Giribaldi G, O'Brien E, Turrini F, Mannu F, Ulliers D, Simula G, Luzzatto L, Arese P (1998) Early phagocytosis of glucose-6-phosphate dehydrogenase (G6PD)-deficient erythrocytes parasitized by Plasmodium falciparum may explain malaria protection in G6PD deficiency. Blood 92:2527–2534

Fernandez-Reyes D, Craig AG, Kyes SA, Peshu N, Snow RW, Berendt AR, Marsh K, Newbold CI (1997) A high frequency African coding polymorphism in the N-terminal domain of ICAM-1 predisposing to cerebral malaria in Kenya. Human Mol Genet 6:1357–1360

Friedman MJ (1978) Erythrocytic mechanism of sickle cell resistance to malaria. Proc Natl Acad Sci USA 75:1994–1997

Golenser J, Chevion M (1989) Oxidant stress and malaria: host-parasite relationships in normal and abnormal erythrocytes. Sem Hematol 26:313–325

Hadley T, Saul A, Lamont G, Hudson DE, Miller LH, Kidson C (1983) Resistance of Melanesian elliptocytes (ovalocytes) to invasion by Plasmodium knowlesi and Plasmodium falciparum malaria parasites in vitro. J Clin Invest 71:780–782

Haldane JBS (1949)Disease and evolution. Ricerca Sci 19 (Suppl. I): 68–76

Hill AV (1998) The immunogenetics of human infectious diseases. Ann Rev Immunol 16:593–617

Hill AV, Elvin J, Willis AC, Aidoo M, Allsopp CE, Gotch FM, Gao XM, Takiguchi M, Greenwood BM, Townsend AR (1992) Molecular analysis of the association of HLA-B53 and resistance to severe malaria [see comments]. Nature 360:434–439

Holt M, Hogan PF, Nurse GT (1981) The ovalocytosis polymorphism on the western border of Papua New Guinea. Human Biol 53:23–34

Jackson (1997) Ecological modeling of human-plant parasite coevolutionary triads: theoretical perspectives on the interrelationships of Human HbβS, G6PD, *Manihot esculenta, Vicia faba,* and *Plasmodium falciparum.* In: Greene LS, Danubio ME (eds) Adaptation to malaria: the interaction of biology and culture. Amsterdam, Gordon & Breach, pp 177–207

Joyce S, Woods AS, Yewdell JW, Bennink JR, De Silva AD, Boesteanu A, Balk SP, Cotter RJ, Brutkiewicz RR (1998) Natural ligand of mouse CD1d1: cellular glycosylphosphatidylinositol. Science 279:1541–1544

Kidson C, Lamont G, Saul A, Nurse GT (1981) Ovalocytic erythrocytes from Melanesians are resistant to invasion by malaria parasites in culture. Proc Natl Acad USA Sci 78:5829–5832

Kurdi-Haidar B, Luzzatto L (1990) Expression and characterization of glucose 6-phosphate dehydrogenase of Plasmodium falciparum. Mol Biochem Parasit 41:83–92

Liu SC, Palek J, Yi SJ, Nichols PE, Derick LH, Chiou SS, Amato D, Corbett JD, Cho MR, Golan DE (1995) Molecular basis of altered red blood cell membrane properties in Southeast Asian ovalocytosis: role of the mutant band 3 protein in band 3 oligomerization and retention by the membrane skeleton. Blood 86:349–358

Luzzatto L (1979) Genetics of red cells and susceptibility to malaria. Blood 54:961–976

Luzzatto L (1981) Genetics of human red cells and susceptibility to malaria. In: Michal F (ed) Modern genetic concepts and techniques in the study of parasites. Basel, Schwabe & Co., pp 257–277

Luzzatto L (1985) Malaria and the red cell. In: Hoffbrand AV (ed) Recent Advances in haematology. 4th ed. London, Churchill Livingstone, pp 109–126

Luzzatto L, Pinching AJ (1990) Commentary to R Nagel – Innate resistance to malaria: the intraerythrocytic cycle. Blood Cells 16:340–347

Luzzatto L, Mehta A (1995) Glucose 6-phosphate dehydrogenase deficiency. In: Scriver CR, Beaudet AL, Sly WS, Valle D (eds) The metabolic and molecular bases of inherited disease. 7th ed. New York, McGraw-Hill, pp 3367–3398

Luzzatto L, Usanga EA, Reddy S (1969) Glucose 6-phosphate dehydrogenase deficient red cells: resistance to infection by malarial parasites. Science 164:839–842

Luzzatto L, Nwachuku-Jarrett ES, Reddy S (1970) Increased sickling of parasitised erythrocytes as mechanism of resistance against malaria in the sickle-cell trait. Lancet i:319–321

Luzzi GA, Merry AH, Newbold CJ, Marsh K, Pasvol G, Weatherall DJ (1991) Surface antigen expression on *Plasmodium falciparum* infected erythrocytes is modified in α- and β-thalassemia. J Exp Med 173:785–791

Maugh ITH (1981) A new understanding of sickle cell emerges. X-ray, kinetic studies paint a comprehensive picture of sickle cell disease to the level of atomic interactions. Science 211:265–267

McGuire W, Knight JC, Hill AV, Allsopp CE, Greenwood BM, Kwiatkowski D (1999) Severe malarial anemia and cerebral malaria are associated with different tumor necrosis factor promoter alleles. J Infect Dis 179:287–290

Modiano G, Morpurgo G, Terrenato L, Novelletto A, Di Rienzo A, Colombo B, Purpura M, Mariani M, Santachiara-Benerecetti S, Brega A, Dixit KA, Shrestha SL, Lania A, Wanachiwanawin W, Luzzatto L (1991) Protection against malaria morbidity: near-fixation of the α-thalassaemia gene in a Nepalese population. Am J Human Genet 48:390–397

Montanaro V, Casamassimi A, D'Urso M, Yoon J-Y, Freije W, Schlessinger D, Muenke M, Nussbaum RL, Saccone S, Maugeri S, Santoro AM, Motta S, Della Valle G (1991) In situ hybridization to cytogenetic bands of yeast artificial chromosomes covering 50 % of human Xq24-Xq28 DNA. Am J Human Genet 48:183–194

Motulsky AG (1960) Metabolic polymorphisms and the role of infectious diseases in human evolution. Human Biol 32:28–62

Newton CR, Warrell DA (1998) Neurological manifestations of falciparum malaria. Ann Neurol 43:695–702

Ntoumi F, Rogier C, Dieye A, Trape JF, Millet P, Mercereau-Puijalon O (1997). Imbalanced distribution of Plasmodium falciparum MSP-1 genotypes related to sickle-cell trait. Mol Med 3:581–592

Nurse GT, Coetzer TL, Palek J (1992) The elliptocytoses, ovalocytosis and related disorders. Baillière's Clin Haematol 5:187–207

O'Brien E, Kurdi-Haidar B, Wanachiwanawin W, Carvajal JL, Villiamy TJ, Cappadoro M, Mason PJ, Luzzatto L (1994) Cloning of the glucose 6-phosphate dehydrogenase gene from Plasmodium falciparum. Mol Biochem Parasit 64:313–326

O'Donnell A, Allen SJ, Mgone CS, Martinson JJ, Clegg JB, Weatherall DJ (1998) Red cell morphology and malaria anaemia in children with Southeast-Asian ovalocytosis band 3 in Papua New Guinea. Br J Haematol 101:407–412

Ohta N, Iwaki K, Itoh M, Fu J, Nakashima S, Hato M, Tolle R, Bujard H, Saitoh A, Tanabe K (1997) Epitope analysis of human T-cell response to MSP-1 of Plasmodium falciparum in malaria-nonexposed individuals. Int Arch Allergy Immunol 114:15–22

Oppenheim A, Jury CL, Rund D, Vulliamy TJ, Luzzatto L (1993) G6PD Mediterranean accounts for the high prevalence of G6PD deficiency in Kurdish Jews. Human Genet 91:293–294

Plebanski M, Aidoo M, Whittle HC, Hill AV (1997) Precursor frequency analysis of cytotoxic T lymphocytes to pre-erythrocytic antigens of Plasmodium falciparum in West Africa. J Immunol 158:2849–2855

Roth JEF, Friedman M, Ueda Y, Tellez I, Trager W, Nagel RL (1978) Sickling rates of human AS red cells infected in vitro with plasmodium falciparum malaria. Science 202:650–652

Roth EF, Jr., Raventos-Suarez C, Rinaldi A, Nagel RL (1983) Glucose 6-phosphate dehydrogenase deficiency inhibits in vitro growth of Plasmodium falciparum. Proc Natl Acad Sci USA 80:298–301

Rowe JA, Moulds JM, Newbold CI, Miller LH (1997) P. falciparum rosetting mediated by a parasite-variant erythrocyte membrane protein and complement-receptor 1. Nature 388:292–295

Ruwende C, Khoo SC, Snow RW, Yates SN, Kwiatkowski D, Gupta S, Warn P, Allsopp CE, Gilbert SC, Peschu N, Newbold CI, Greenwood BM, Marsh K, Hill AVS (1995) Natural selection of hemi- and heterozygotes for G6PD deficiency in Africa by resistance to severe malaria. Nature 376:246–249

Schofield AE, Reardon DM, Tanner MJ (1992) Defective anion transport activity of the abnormal band 3 in hereditary ovalocytic red blood cells. Nature 355:836–838

Schofield L, McConville MJ, Hansen D, Campbell AS, Fraser-Reid B, Grusby MJ, Tachado SD (1999) CD1d-restricted immunoglobulin G formation to GPI-anchored antigens mediated by NKT cells. Science 283:225–229

Shear HL, Roth EF, Fabry ME, Costantini FD, Pachnis A, Hood A, Nagel RL (1993) Transgenic mice expressing human sickle hemoglobin are partially resistant to rodent malaria. Blood 81:222–226

Siniscalco M, Bernini L, Filippi G, Latte B, Khan PM, Piomelli S, Rattazzi M (1966) Population genetics of haemoglobin variants, thalassaemia and glucose-6-phosphate dehydrogenase deficiency, with particular reference to the malaria hypothesis. Bull WHO 34:379–393

Sultan AA, Thathy V, Frevert U, Robson KJ, Crisanti A, Nussenzweig V, Nussenzweig RS, Menard R (1997) TRAP is necessary for gliding motility and infectivity of plasmodium sporozoites. Cell 90:511–522

Terrenato L, Shrestha S, Dixit M, Luzzatto L, Modiano G, Morpurgo G, Arese P (1988) Decreased malaria morbidity in the Tharu people compared to sympatric populations in Nepal. Ann Trop Med Parasitol 82:1–11

Trager W, Jensen JB (1976) Human malaria parasites in continuous culture. Science 193:673–675

Usanga EA, Luzzatto L (1985) Adaptation of Plasmodium falciparum to glucose 6-phosphate dehydrogenase deficient host red cells by production of parasite-encoded enzyme. Nature 313:793–795

Vulliamy T, Luzzatto L, Hirono A, Beutler E (1997) Hematologically important mutations: glucose 6-phosphate dehydrogenase. Blood Cells Mol Dis 23:292–303

Weatherall DJ, Abdalla S, Pippard MJ (1983) The anaemia of Plasmodium falciparum malaria. In: Evered D, Whelan J (eds) Malaria and the red cell. London, Pitman, pp 74–97

Williams TN, Maitland K, Bennett S, Granzakowski M, Peto TE, Newbold CI, Bowden DK, Weatherall DJ, Clegg JB (1996) High incidence of malaria in alpha-thalassaemic children [see comments]. Nature 383:522–525

Genetic Basis of Resistance to Alzheimer's Disease and Related Neurodegenerative Diseases

C. L. Masters and K. Beyreuther

Summary

There are two classes of adverse risk factors for Alzheimer's disease(AD): increasing age and genetics. If longevity itself is determined to a large extent by genetic factors, then the scene is set for a complex set of interactions determining the relative balance between genetic susceptibility and resistance. The first major task is to clarify whether the exponential rise in incidence of AD between the ages of 40 and 90 continues in the very elderly (ages 90–110). With the advent of more accurate analyses of the cerebral plaque (amyloid Aβ) and tangle (tau) loads, this question can now be solved. If it is found that the incidence of AD does in fact decrease (as some preliminary studies suggest), then "longevity genes" may act through an AD-protective pathway.

Currently, there are three known genes in which mutations cause AD, usually with an age of onset under 65 years. These mutant genes, amyloid precursor protein (APP) and presenilin 1 and 2 (PS1,2), act through the amyloid Aβ pathway, perhaps by altering the efficiencies with which APP is processed by the α, β and γ secretases. The accumulation of amyloid Aβ in the brain is then thought to result in a toxic gain of function, the result of which is neurodegeneration. In contrast, there are genetic risk factors [the best known are the alleles of apolipoprotein E (ApoE)] that may act through the APP/Aβ or some other pathway. The ApoE phenomenon is most instructive and probably serves as a model for other, as yet undefined, genetic risk factors.

There are three common alleles of ApoE (designated ε2, 3, 4). Experimentally, it has been determined that the complete absence of ApoE is beneficial, in that APP transgenic mice on the ApoE$^{-/-}$ background fail to develop Aβ amyloid plaques. In humans, the ApoE ε4 allele is associated, in a dose-dependant fashion, with an earlier onset of AD. Conversely, the ε2 allele may even confer some degree of resistance to, or delay the onset of, AD. There is accumulating evidence that the ApoE effect acts on the APP/Aβ pathway, either through a direct interaction with Aβ in the extracellular space or through the modulation of cholesterol metabolism or the LRP receptor.

Other candidate genetic risk factors are beginning to emerge. With more sophisticated genetic-association strategies, α$_2$ macroglobulin (which may also clear Aβ through the LRP receptor) has been proposed as a major AD determinant.

V. Boulyjenkov, K. Berg, Y. Christen (Eds.)
Genes and Resistance to Diseases
© Springer-Verlag Berlin Heidelberg 2000

For each genetic risk factor thus identified, there will be a "resistance" and "susceptibility" aspect to their modes of action. The future, therefore, looks particularly promising for the beneficial application of this knowledge, in both a diagnostic/prognostic sense and for the development of new therapeutic strategies.

Introduction

The emerging picture of neurodegenerative disorders suggests a common process in which a normal cellular protein undergoes a series of conformational changes, resulting in a toxic gain of function associated with the accumulation of relatively insoluble aggregates of the protein. For most of the major neurodegenerative conditions, there is now a separately identifiable gene product that lies at the center of the pathogenic pathway (Table 1). While there is still much to be learned about the detailed molecular basis of this toxic gain of function, the realisation of this common process opens many new therapeutic opportunities.

For each neurodegenerative disease, therefore, there is a distinct biochemical pathway that leads to the toxic end-product. Multiple genetic, epigenetic and environmental factors can interact in this pathway. These factors include all aspects of susceptibility and resistance, of which very little is currently known. Relatively few causative genes have been identified, but each so far has made a major impact in our understanding of the disease process. In contrast, there is no clear evidence of a causative environmental factor [unless one accepts the modes of transmissibility of Creutzfeldt-Jakob disease (CJD) as falling into this category], but strong indirect evidence for their existence could be inferred from the epidemiology of conditions like the Guamanian Parkinsonism-dementia-amyotrophic lateral sclerosis (ALS) complex.

In this chapter, we use the paradigm of Alzheimer's disease (AD) as the best-studied model of a neurodegenerative condition, and explore the possibility that certain genetic resistance factors exist. In choosing this particular disease, one is immediately faced with the task of distinguishing the concepts of "absence of AD" from "normal biological longevity." Elsewhere in this symposium, Dr. F. Schächter discusses the properties of genes that determine longevity. Since we take the position that AD is a distinct pathological process, and not an inherent part of the "normal" biological aging process, we would consider that any AD-specific resistance factor should not be classified within those genes that determine the normal biological life span (for a recent review on this subject, see Finch and Tanzi 1997).

The Molecular Pathology of AD

The principal molecular players in Alzheimer's neurodegeneration are the Alzheimer Aβ amyloid precursor protein (AAP), the presenilins (PS1,2), apolipoprotein E (ApoE), and the microtubule associated protein tau. Many other candi-

Table 1. The major neurodegenerative diseases associated with a faulty gene product

	Product	Causative genes	Genetic susceptibility		Genetic resistance
Alzheimer's disease (AD)[a]	APP/Aβ	APP PS1,2	ApoE ε4 α2M LRP	↔ ↔ ↔	ApoE ε2 ? ?
Creutzfeldt-Jakob Disease (CJD)[b]	PrP	PRNP	Codon 129 of PRNP Homozygosity		Heterozygosity
Parkinson's disease (PD)[c]	αSN	αSN	?		?
Motor neuron disease (ALS)[d]	SOD1	SOD1	?		?
Huntington's disease (HD)[e]	polyglutamine	Htt	Extent of poly-Q expansion		?
Fronto temporal dementia and Parkinsonism (FTDP-17)[f]	4-repeat tau	tau	?		?

[a] AD, Alzheimer's disease; APP, the Alzheimer amyloid precursor protein; Aβ, Alzheimer amyloid; PS1,2, presenilin 1,2; ApoE, apolipoprotein E, ε2 or 4 allotypes; α2M, alpha 2 macroglobulin; LRP, lipoprotein related protein receptor. Related variants in this category include amyloid congophilic angiopathy.

[b] CJD, Creutzfeldt-Jakob disease; PrP, prion protein; PRNP, gene encoding PrP. Also included are the Gerstmann-Sträussler-Scheinker syndrome (GSS), fatal familial insomnia (FFI) and the new variant form of CJD related to bovine spongiform encephalopathy (BSE) or mad cow disease.

[c] PD, Parkinson's disease; αSN, alpha synuclein. Also included in this category are Diffuse Lewy Body Disease (DLBD) and multiple system atrophy (MSA).

[d] ALS, amyotrophic lateral sclerosis or motor neuron disease; SOD1, Cu Zn superoxide dismutase 1. Conditions like the Guamanian forms of ALS with Parkinson's disease (ALS/PD) may prove to be related.

[e] HD, Huntington's disease; Htt, Huntington protein. All conditions in which there is expression of an unstable polyglutamine (poly Q) tract may be classed in this group, including the various spinocerebellar ataxias and Machado-Joseph disease (MJD).

[f] FTDP-17, frontotemporal dementia with Parkinsonism linked to chromosome 17; also referred to as a "tauopathy." This may include forms of Pick's lobar atrophy (PLA), familial non-specific dementia (Kraepelin's disease), some forms of non-specific dementia associated with ALS (so-called amyotrophic forms of CJD) and the frontal lobe degenerations.

dates exist, and undoubtedly many others remain to be identified. Nevertheless, the processing of APP provides the central axis of the molecular pathology of AD.

The structural domains and features of APP (Fig. 1) show a type 1 integral transmembrane protein that may play a role in cellular membrane stabilization, synaptic modulation, and neurite outgrowth (Storey et al. 1996 a,b; Williamson et al. 1996; Morimoto et al. 1998; Koizumi et al. 1998; Meziane et al. 1998). The N-terminal 100 residues have now been resolved by x-ray crystallography and disclose a growth factor-like folding region in conjunction with the first predicted heparin binding domain (HBD-1; Rossjohn et al. 1999). This region of APP is

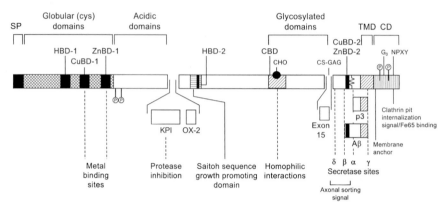

Fig. 1. Structural domains of the Alzheimer Aβ amyloid precursor protein. Abbreviations: SP, signal peptide; HBD, heparan binding domain; Zn/Cu BD, zinc and copper binding sites; CBD, collagen binding domain; CS-GAG, chondroitin sulfate glycosaminoglycan site; TMD, transmembrane domain; CD, cytoplasmic domain; Go, G-protein interaction; KPI, Kunitz protease inhibitor.

also active in the stabilisation of platelet activation and adhesion (Henry et al. 1998) and may play a role in the anti-integrin phenotype of APP over-expression in *Drosophila* (Fossgreen et al. 1998). The identification of the metal-interaction sites (Hesse et al. 1994; Multhaup et al. 1995, 1996, 1998) led directly to the examination of the role of metals in the aggregation and redox state and toxicity of the Aβ domain (Huang et al. 1999; White et al. 1998, 1999; Fu et al. 1998). The function of the alternatively spliced protease inhibitory domain remains enigmatic (Moir et al. 1998). Splicing of exon 15 near the transmembrane domain directly controls the chondroitin-sulfation site (Hartmann et al. 1996) and may indirectly affect the axonal targeting of APP (Tienari et al. 1996). Despite considerable effort, it has been difficult to ascribe an activity to any region of the ectodomain of APP which could directly affect the critical processing events that are responsible for the generation of the Aβ product. Definition of the sites of APP-PS1, 2 interactions (see below) may prove to be relevant. Stabilisation of the transmembrane domain through the triple lysine membrane anchor could be expected to affect Aβ biogenesis, and some recent evidence indicates that a variety of interacting proteins binding to the cytoplasmic domain or other regions of APP may affect the half life of APP (and thereby influence its residence time in cellular compartments responsible for Aβ generation; Yang et al. 1998; Borg et al. 1998; Duilio et al. 1998).

Looking in more detail at the events that surround the biogenesis of Aβ amyloid (Fig. 2), it is seen that the products that accumulate in the AD brain – amyloid plaques (APC), perivascular amyloid (ACA) – have very ragged N- and C-termini. Variations in length at the hydrophobic C-terminus (Aβ40, Aβ42) are now thought to be important determinants of β-sheet content, aggregability and neurotoxicity: the longer Aβ42/43 forms possess more of these undesirable properties. The smaller P3 product (Aβ17 – 40/42) is also identifiable within the amy-

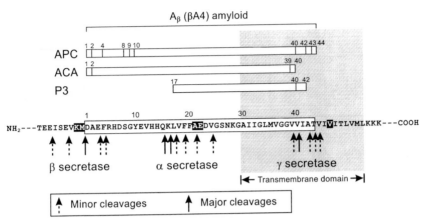

Fig. 2. The major proteolytic products of the Aβ domain. APC, amyloid plaque core; ACA, amyloid congophilic angiopathy; P3, the 3kd α-secretase product. The pathogenic mutations are shown highlighted near the respective secretase sites.

loid deposits, but the degree to which it participates in the neurodegenerative process remains uncertain.

The proteolytic events leading to the release of the APP ectodomain from the cell membrane (the α- and β-secretase activities) are becoming clearer with the recent cloning of the TACE-class of membrane-associated metalloproteases (Buxbaum et al. 1998) and the angiotensin-converting enzyme secretase (Parvathy et al. 1998). It appears that they are good candidates for α-secretase activity. The β-secretases and γ-secretases are being identified, and much progress is being made in defining the cellular compartments in which they operate (Fuller et al. 1995; Ikezu et al. 1998; Tienari et al. 1997; Peraus et al. 1997; Hartmann et al. 1997; Li et al. 1999). A view is emerging in which the generation of Aβ42 occurs in a pre-Golgi compartment, whereas Aβ40 is generated between the Golgi and cell surface. The rare mutations in the APP gene which cause early onset and aggressive forms of AD cluster around the secretase sites: the "Swedish" mutations at the β-secretase site increase the total amount of Aβ released from cell; the "London" mutations around the γ-secretase sites have the effect of producing a higher ratio of Aβ42:40 (Lichtenthaler et al. 1997, 1999). The "Dutch/Flemish" mutations near the α-secretase sites affect either the solubility of Aβ or presumptively lead to a relative resistance to α-secretase action (de Jonghe et al. 1998).

The more common presenilin mutations, which also result in early-onset aggressive AD, all appear to operate at the level of γ-secretase activity, resulting in a higher ratio of Aβ42:40. The mechanisms underlying this phenomenon are now being uncovered (Weidemann et al. 1997; Culvenor et al. 1997; Scheuner et al. 1996). Importantly, loss of PS activity has a major down-regulatory effect on γ-secretase, with the accumulation of C-terminal fragments of APP (De Strooper et al. 1998).

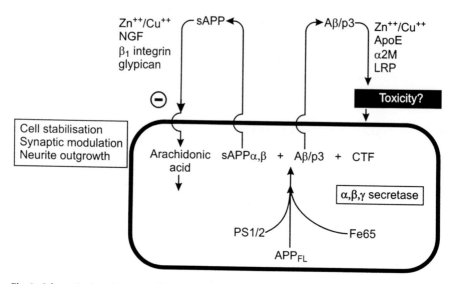

Fig. 3. Schematic depiction of a cell, with the interactions of full-length APP, presenilins and cytoplasmic interacting factors (Fe65) yielding the major secreted APP (sAPPα,β) amyloidogenic Aβ/P3, and C-terminal fragments (CTF). Release of soluble APP from the cell allows it to act back on the cell through an unidentified receptor. This action is inhibitory and involves an arachidonic acid-mediated pathway. Aβ/P3 exerts a toxic effect on the cell, with modulation by metals ions (Zn⁺⁺/Cu⁺⁺), binding proteins (ApoE, α2M) or receptors like LRP.

As an integrated system (Fig. 3), the metabolism of APP and its products allows for multiple points at which genetic, epigenetic and environmental factors can interact. The most important variable, however, may be the absolute level of Aβ in a "soluble" phase, either intracellularly or extracellulary (Fig. 4; McLean et al. 1999). A clue to this process may lie in the interactions between ApoE and Aβ (Russo et al. 1998), through which the age at onset of sporadic AD may be related (Strittmater et al. 1993).

There is now a vast compendium of publications on the adverse risks associated with the ε4 allele of ApoE, but much less is appreciated on the possible protective effects of the ε2 allele. The complete absence of ApoE in the murine transgenic model of AD causes a diminution in the visible Aβ amyloid burden (Bales et al. 1997). This finding would be consistent with the concept that the physically tight associations between ApoE and Aβ may act to alter the degradation and clearance of Aβ from the extracellular compartment of the brain interstitium. To what degree this interaction modulates any toxic activity of Aβ remains to be clarified. Alternate hypotheses on the action of ApoE involve its role in cellular cholesterol metabolism, in which it has recently been demonstrated that cholesterol depletion has a major effect on α- or β-secretase activity (Bodovitz and Klein 1996; Simons et al. 1998). As a major transport mechanism for delivering cholesterol to the brain, ApoE could easily affect the overall cerebral balance of cholesterol homeostasis.

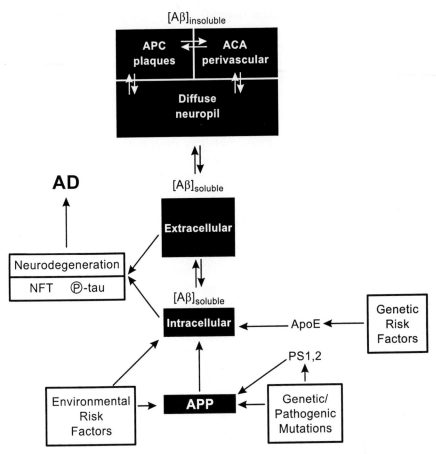

Fig. 4. The release of soluble Aβ from an intracellular to an extracellular pool may be a principal determinant of the severity of AD. The insoluble Aβ pool (locked in amyloid plaques (APC) and perivascular amyloid (ACA) is unlikely to be of consequence to the integrity of the surrounding neurons. Neurodegeneration, associated with hyperphosphorylated tau in neurofibrillary tangles, is seen as a direct consequence of the effects of soluble Aβ.

Finally, the pathway that leads to paired helical filaments and neurofibrillary tangles is still very obscure, despite the demonstration that the microtubule associated protein tau is a major constituent (Goedert et al. 1996). The general concept that is arising from the discovery of mutant tau causing frontotemporal dementia with Parkinsonism (FTDP-17) is that a failure of proper binding of tau to tubulin may leave an intracellular pool of free tau, which is then susceptible to aggregation and polymerisation (Hong et al. 1998). Hyperphosphorylation of this pool of tau may also contribute to this process. This still leaves the connection between Aβ and tau-PHF unexplained. One attractive hypothesis is that the oxidative stress generated by Aβ itself can induce tau to form aggregates and fibrils.

Pharmacogenetics and Protection from AD

With the advent of large-scale genomic probes, it is highly likely that the concept of "genetic resistance" will rapidly evolve to include the genomic profiles of individuals who respond favourably or adversely to drugs directed at AD. It is to be hoped that the dentification of these "genetically responsive" individuals will lead to a clearer understanding of the genes involved in promoting the beneficial actions of the drugs that are emerging for the treatment of AD. Favourable responses to the acetylcholinesterase inhibitors are currently providing the testing grounds for these concepts. In the near future, we can look forward to the analysis of responses to γ-secretase inhibitors and perhaps anti-oxidants.

Evolutionary Pressures and Resistance to AD

What selective advantages could exist which would allow AD to be such a prominent feature of the aging, post-reproductive, human brain? Surveying the phylogenetic spectrum of APP and the APLP superfamily, it is clear that this highly conserved family of genes must subserve an important series of normal functions (von Koch et al. 1997). Although the processing machinery exists in the yeast, the APP homologue does not (Le Broque et al. 1998). The most "primitive" species with APP appears to be *Drosophila*, in which the human APP gene can rescue a mutant phenotype (Luo et al. 1992).

Not unexpectedly, non-human primates show very high homology with human APP, particularly in the critical juxta- and trans-membrane domain areas concerned with Aβ biogenesis. The Old World primates develop lesions that closely approximate those of AD, yet none have been described with a very aggressive disease process that clearly causes a widespread neurodegenerative change of severity comparable to the typical human case. To an even lesser extent, aged dogs and bears also develop Aβ plaques. Importantly, aged rodents (mice and rats) do not develop Aβ plaques or tau-NFT. Recent experiments with the rodent Aβ sequence (in which there are three amino acids that differ from the human) show that, both in vitro and in vivo, the rodent Aβ has less inherent toxicity and a lower capacity for metal-induced aggregation, and appears unable per se to act as a substrate for Aβ plaque formation in transgenic animals (Atwood et al. 1998). This finding raises fundamental questions about the normal function of APP in general and about the processing events that serve to release Aβ. It has been speculated that Aβ itself may have a normal activity that in some manner contributes to the overall function of APP. If this activity of Aβ could be related to a selective advantage in the pre-reproductive phase of human development and aging, then an explanation for its perceived side-effects (the toxic principle underlying Alzheimerization of the brain) might be forthcoming. A corollary of this line of speculation is that any natural resistance genes to AD should not have been under any selective pressure. If this were correct, then the search for genes that confer resistance to AD may prove fruitless.

To end on a more optimistic note, it remains possible that serendipitous (i.e., without selective advantage) interactions been genes and the APP/Aβ pathway may yet emerge. One example could be the possible protective effect of diabetes (hyperglycemia or hypoinsulinemia) on the risk of developing AD. The complex genetic predisposition to late-onset diabetes may therefore work against the risks of developing AD. Much genetic epidemiology remains to be performed to tease out the web of interactions in these complex degenerative disorders.

References

Atwood CS, Moir RD, Huang XD, Scarpa RC, Bacarra NME, Romano DM, Hartshorn MK, Tanzi RE, Bush AI (1998) Dramatic aggregation of Alzheimer Aβ by Cu(II) is induced by conditions representing physiological acidosis. J Biol Chem 273:12817–12826

Bales KR, Verina T, Dodel RC, Du YS, Altstiel L, Bender M, Hyslop P, Johnstone EM, Little SP, Cummins DJ, Piccardo P, Ghetti B, Paul SM (1997) Lack of apolipoprotein E dramatically reduces amyloid β-peptide deposition. Nature Genet 17:263–264

Bodovitz S, Klein WL (1996) Cholesterol modulates alpha-secretase cleavage of amyloid precursor protein. J Biol Chem 271:4436–4440

Borg JP, Yang YN, Detaddéo-Borg M, Margolis B, Turner RS (1998) The X11α: protein slows cellular amyloid precursor protein processing and reduces Aβ40 and Aβ42 secretion. J Biol Chem 273:14761–14766

Buxbaum ID, Liu KN, Luo YX, Slack JL, Stocking KL, Peschon JJ, Johnson RS, Castner BJ, Cerretti DP, Black RA (1998) Evidence that tumor necrosis factor alpha converting enzyme is involved in regulated alpha-secretase cleavage of the Alzheimer amyloid protein precursor. J Biol Chem 273:27765–27767

Culvenor JC, Maher F, Evin G, Malchiodi-Albedi F, Cappai R, Underwood JR, Davis JB, Karran EH, Roberts GW, Beyreuther K, Masters CL (1997) Alzheimer's disease-associated presenilin 1 in neuronal cells: Evidence for localization to the endoplasmic reticulum-Golgi intermediate compartment. J Neurosci Res 49:719–731

De Jonghe C, Zehr C, Yager D, Prada CM, Younkin S, Hendriks L, van Broeckhoven C, Eckman CB (1998) Flemish and Dutch mutations in amyloid β precursor protein have different effects on amyloid β secretion. Neurobiol Disease 5:281–286

De Strooper B, Saftig P, Craessaerts K, Vanderstichele H, Guhde G, Annaert W, Von Figura K, Van Leuven F (1998) Deficiency of presenilin-1 inhibits the normal cleavage of amyloid precursor protein. Nature 391:387–390

Duilio A, Faraonio R, Minopoli G, Zambrano N, Russo T (1998) Fe65L2 – a new member of the Fe65 protein family interacting with the intracellular domain of the Alzheimers β-amyloid precursor protein. Biochem J 330:513–519

Finch CE, Tanzi RE (1997) Genetics of aging. Science 278:407–411

Fossgreen A, Brückner B, Czech C, Masters CL, Beyreuther K (1998) Transgenic expressing human amyloid precursor protein show γ-secretase activity and a blistered-wing phenotype. Proc Natl Acad Sci USA 95:13703–13708

Fu WM, Luo H, Parthasarathy S, Mattson MP (1998) A Catecholamines potentiate amyloid β-peptide neurotoxicity – involvement of oxidative stress, mitochondrial dysfunction, and perturbed calcium homeostasis. Neurobiol Disease 5:229–243

Fuller SJ, Storey E, Li Q-X, Smith AI, Beyreuther K, Masters CL (1995) Intracellular production of βA4 amyloid of Alzheimer's disease: modulation by phosphoramidon and lack of coupling to the secretion of the amyloid precursor protein. Biochemistry 34:8091–8098

Goedert M, Jakes R, Spillantini MC, Hasegawa M, Smith MJ, Crowther RA (1996) Assembly of microtubule-associated protein tau into Alzheimer-like filaments induced by sulphated glycosaminoglycans. Nature 383:550–553

Hartmann T, Bergsdorf C, Sandbrink R, Tienari PJ, Multhaup G, Ida N, Bieger S, Dyrks T, Weidemann A, Masters CL, Beyreuther K (1996) Alzheimer's disease βA4 protein release and APP sorting are regulated by alternative splicing. J Biol Chem 271:13208–13214

Hartmann T, Bieger SC, Brühl B, Tienari PJ, Ida N, Allsop D, Roberts GW, Masters CL, Dotti CG, Unsicker K, Beyreuther K (1997) Distinct sites of intracellular production for Alzheimer's disease Aβ40/42 amyloid peptides. Nature Med 3:1016–1020

Henry A, Li QX, Galatis D, Hesse L, Multhaup G, Beyreuther K, Masters CL, Cappai R (1998) Inhibition of platelet activation by the Alzheimer's disease amyloid precursor protein. Brit J Haematol 103:402–415

Hesse L, Beher D, Masters CL, Multhaup G (1994) The βA4 amyloid precursor protein binding to copper. FEBS Lett 349:109–116

Hong M, Zhukareva V, Vogelsberg-Ragaglia V, Wszolek Z, Reed L, Miller BI, Geschwind DH, Bird TD, McKeel D, Goate A, Morris JC, Wilhelmsen KC, Schellenberg GD, Trojanowski JQ, Lee VMY (1998) Mutation-specific functional impairments in distinct tau isoforms of hereditary FTDP-17. Science 282:1914–1917

Huang X, Cuajungco MP, Atwood CS, Hartshorn MA, Tyndall J, Hanson GR, Stokes KC, Leopold M, Multhaup G, Lee E, Goldstein LE, Scarpa RC, Saunders AJ, Lim J, Moir RD, Glabe C, Bowden EF, Masters CL, Fairlie DP, Tanzi RE, Bush AI (1999) Cu(II) potentiation of Alzheimer Aβ neurotoxicity. Correlation with cell-free hydrogen peroxide production and metal reduction. J Biol Chem 274:37111–37116

Ikezu T, Trapp BD, Song KS, Schlegel A, Lisanti MP, Okamoto T (1998) Caveolae, plasma membrane microdomains for alpha-secretase-mediated processing of the amyloid precursor protein. J Biol Chem 273:10485–10495

Koizumi S, Ishiguro M, Ohsawa I, Morimoto T, Takamura T, Inoue K, Kohsaka S (1998) The effect of a secreted form of β-amyloid-precursor protein on intracellular Ca^{2+} increase in rat cultured hippocampal neurones. Brit J Pharmacol 123:1483–1489

Le Brocque D, Henry A, Cappai R, Li Q-X, Tanner JE, Galatis D, Cray C, Holmes S, Underwood JR, Beyreuther K, Masters CL, Evin G (1998) Processing of the Alzheimer's disease amyloid precursor protein in Pichia pastoris: α-, β- and γ-secretase products. Biochemistry 37:14958–14965

Li Q-X, Maynard C, Cappai R, McLean CA, Cherny RA, Lynch T, Culvenor JG, Trevaskis J, Tanner JE, Bailey KA, Czech C, Bush AJ, Beyreuther K, Masters CL (1999) Intracellular accumulation of detergent-soluble amyloidogenic Aβ fragment of Alzheimer's disease precursor protein in hippocampus of aged transgenic mice. J Neurochem 72:2479–2487

Lichtenthaler SF, Ida N, Multhaup G, Masters CL, Beyreuther K (1997) Mutations in the transmembrane domain of APP altering γ-secretase specificity. Biochemistry 36:15396–15403

Lichtenthaler SF, Wang R, Masters CL, Beyreuther K (1999) Mechanism of the cleavage specificity of Alzheimer's disease γ-secretase revealed by phenylalanine-scanning mutagenesis. Proc Natl Acad Sci USA, in press

Luo L, Tully T, White K (1992) Human amyloid precursor protein ameliorates behavioural deficit of flies deleted for Appl gene. Neuron 9:595–605

McLean CA, Cherney RA, Fraser FW, Fuller SJ, Smith MJ, Beyreuther K, Bush AI, Masters CL (1999) Soluble pool of Aβ as a determinant of severity of neurodegeneration in Alzheimer's disease. Ann Neurol 46:860–866

Meziane H, Dodart JC, Mathis C, Little S, Clemens J, Paul SM, Ungerer A (1998) Memory-enhancing effects of secreted forms of the β-amyloid precursor protein in normal and amnestic mice. Proc Natl Acad Sci USA 95:12683–12688

Moir RD, Lynch T, Bush AI, Whyte S, Henry A, Portbury S, Multhaup G, Small DH, Tanzi RE, Beyreuther K, Masters CL (1998) Relative increase in Alzheimer's disease of soluble forms of cerebral Aβ amyloid protein precursor containing the Kunitz protease inhibitory domain. J Biol Chem 273:5013–5019

Morimoto T, Ohsawa I, Takamura C, Ishiguro M, Nakamura Y, Kohsaka S (1998) Novel domain-specific actions of amyloid precursor protein on developing synapses. J Neurosci 18:9386–9393

Multhaup G, Mechler H, Masters CL (1995) Characterization of the high affinity heparin binding site of the Alzheimer's disease βA4 amyloid precursor protein (APP) and its enhancement by zinc (II). J Molec Recog 8:247–257

Multhaup G, Schlicksupp A, Hesse L, Beher D, Ruppert T, Masters CL, Beyreuther K (1996) The amyloid precursor protein of Alzheimer's disease in the reduction of copper (II) to copper (I). Science 271:1406–1409

Multhaup G, Ruppert T, Schlicksupp A, Hesse L, Bill E, Pipkorn R, Masters CL, Beyreuther K (1998) Copper-binding amyloid precursor protein undergoes a site-specific fragmentation in the reduction of hydrogen peroxide. Biochemistry 37:7224–7230

Parvathy S, Karran EH, Turner AJ, Hooper NM (1998) The secretases that cleave angiotensin converting enzyme and the amyloid precursor protein are distinct from tumour necrosis factor-α convertase. FEBS Lett 431:63–65

Peraus G, Masters CL, Beyreuther K (1997) Late compartments of amyloid precursor protein transport in SY5Y cells are involved in β-amyloid secretion. J Neurosci 17:7714–7724

Rossjohn J, Cappai R, Feil SC, Henry A, McKinstry WJ, Galatis D, Hesse L, Multhaup G, Beyreuther K, Masters CL, Parker MW (1999) Crystal structure of the N-terminal, growth factor-like domain of Alzheimer amyloid precursor protein. Nature Struct Biol 6:327–331

Russo C, Angelini G, Dapino D, Piccini A, Piombo G, Schettini G, Chen S, Teller JK, Zaccheo D, Gambetti P, Tabaton M (1998) Opposite poles of apolipoprotein E in normal brains and in Alzheimer's disease. Proc Natl Acad Sci USA 95:15598–15602

Scheuner D, Eckman C, Jensen M, Song X, Citron M, Suzuki N, Bird TD, Hardy J, Hutton M, Kukull W, Larson E, Levy-Lahad E, Viitanen M, Peskind E, Poorkaj P, Schellenberg G, Tanzi R, Wasco W, Lannfelt L, Selkoe D, Younkin S (1996) Secreted amyloid β-protein similar to that in the senile plaques of Alzheimer's disease is increased *in vivo* by the presenilin 1 and 2 and *APP* mutations linked to familial Alzheimer's disease. Nature Med 2:864–870

Simons M, Keller P, DeStrooper B, Beyreuther K, Dotti CG, Simons K (1998) Cholesterol depletion inhibits the generation of β-amyloid in hippocampal neurons. Proc Natl Acad Sci USA 95:6460–6464

Storey E, Spurck T, Pickett-Heaps J, Beyreuther K, Masters CL (1996a) The amyloid precursor protein of Alzheimer's disease is found on the surface of static but not actively motile portions of neurites. Brain Research 735:59–66

Storey E, Beyreuther K, Masters CL (1996b) Alzheimer's disease amyloid precursor protein on the surface of cortical neurons in primary culture co-localizes with adhesion patch components. Brain Res 735:217–231

Strittmatter WJ, Saunders AM, Schmechel D, Pericak-Vance M, Enghild J, Salvesen GS, Roses AD (1993) Apolipoprotein E: high-avidity binding to beta-amyloid and increased frequency of type 4 allele in late-onset familial Alzheimer disease. Proc Natl Acad Sci USA 90:1977–1981

Tienari PJ, De Strooper B, Ikonen E, Simons M, Weidemann A, Czech C, Hartmann T, Ida N, Multhaup G, Masters CL, Van Leuven F, Beyreuther K, Dotti CG (1996) The β-amyloid domain is essential for axonal sorting of amyloid precursor protein. EMBO J 15:5218–5229

Tienari PJ, Ida N, Ikonen E, Simons M, Weidemann A, Multhaup G, Masters CL, Dotti CG, Beyreuther K (1997) Intracellular and secreted Alzheimer β-amyloid species are generated by distinct mechanisms in cultured hippocampal neurons. Proc Natl Acad Sci USA 794:4125–4130

von Koch CS, Zheng H, Chen H, Trumbauer M, Thinakaran G, Van der Ploeg LHT, Price DL, Sisodia SS (1997) Generation of APLP2 KO mice and early postnatal lethality in APLP2/APP double KO mice. Neurobiol Aging 18:661–669

Weidemann A, Paliga K, Dürrwang U, Czech C, Evin G, Masters CL, Beyreuther K (1997) Formation of stable complexes between two Alzheimer's disease gene products: Presenilin-2 and β-amyloid precursor protein. Nature Med 3:328–332

White AR, Zheng H, Galatis D, Maher F, Hesse L, Multhaup G, Beyreuther K, Masters CL, Cappai R (1998) Survival of cultured neurons from amyloid precursor protein knock-out mice against Alzheimer's amyloid-β toxicity and oxidative stress. J Neurosci 18:6207–6217

White AR, Bush A, Beyreuther K, Masters CL, Cappai R (1999) Exacerbation of copper toxicity in primary neuronal cultures depleted of cellular glutathione. J Neurochem 72:2092–2098

Williamson TC, Mok SS, Henry A, Cappai R, Lander AO, Nurcombe V, Beyreuther K, Masters CL, Small DH (1996) Secreted glypican binds to the amyloid precursor protein of Alzheimer's disease (APP) and inhibits APP-induced neurite outgrowth. J Biol Chem 271:31215–31221

Yang Y, Turner RS, Gaut JR (1998) The chaperone BiP/GRP78 binds to amyloid precursor protein and decreases Aβ40 and Aβ42 secretion. J Biol Chem 273:2552–2555

Genes Affecting Cognitive and Emotional Functions

P. McGuffin

Summary

With the exception of comparatively rare single gene disorders causing mental handicap or acquired cognitive decline later in life, the genetic basis of cognitive and emotional function is complex. There is overwhelming evidence from the studies of families, twins and adoptees that cognitive ability, as measured for example by IQ tests, at least 50 % heritable. Similarly, studies of emotionality in twins, as indexed by measures of personality within the normal range or occurrence of clinical depression, show greater similarity in monozygotic than dizygotic pairs, indicating a genetic component. However, the fact that monozygotic twins are not identical with respect to cognitive ability or emotional status demonstrates that environmental factors are also at work. This, together with the likelihood that there are several or perhaps many genes influencing cognition or emotional functions, means that each gene on its own will have a comparatively small effect. Hence, locating and identifying the relevant genes are far more difficult tasks than for single-gene traits or disorders. Although work on laboratory-bred animals has produced some interesting preliminary results in detecting quantitative trait loci (QTLs), rodent models of emotionality or cognitive ability are relatively crude and cannot do justice to the complexity of human behaviours.

Current attempts to identify genes have largely focused on variations at candidate loci, for example, genes relevant to a serotonergic pathways, and have mainly used allelic association, an approach capable of detecting genes of small effect. However, there is increasing interest in attempting whole genome searches looking for linkage disequilibrium. Because linkage disequilibrium is usually only found over small distances (1 centimorgan or less), this approach involves an enormous amount of genotyping. In an attempt to overcome this difficulty we have developed a DNA pooling method as a means of rapidly scanning whole chromosomes. Some promising data using this method on a study of cognitive ability will be discussed.

V. Boulyjenkov, K. Berg, Y. Christen (Eds.)
Genes and Resistance to Diseases
© Springer-Verlag Berlin Heidelberg 2000

Introduction

Simple Mendelian effects are rarely observed in behaviour genetics. Although there are many single gene causes of mental retardation (Thapar and McGuffin 1994), none of these has a population risk of greater than one in a thousand and collectively they account for only a minority of cases of low intelligence. Similarly, although single gene disorders can cause cognitive decline in mid-life, such mutations, for example early onset familial Alzheimer's disease (Masters and Beyreuther, this volume), are again comparatively rare. In some forms of mental retardation and most varieties of pre-senile dementia, emotional changes are seen but there are no known single gene disorders that purely affect emotion, despite many attempts using linkage approaches to detect major autosomal or ex-link genes causing manic depression (McGuffin et al. 1994). Indeed, it seems likely that common behaviours and common abnormalities of behaviour will generally have a multi-factorial basis and, where genes are involved, this will involve the combined action of several, perhaps many, genes, each of which on its own exerts only a small effect (Plomin et al. 1994).

This chapter will focus on two major areas of behaviour, cognitive ability and emotion, as well as some recent work on the genetics of the commonest disorder of emotion, unipolar depression. Molecular genetic studies of cognitive ability and emotion have a relatively recent history, but there is a substantial body of evidence favouring a genetic contribution to such traits based on classic quantitative genetic approaches. This will be briefly reviewed.

Cognitive Function

Although it has often been the subject of controversy and debate (Kamin 1974), the genetic basis of cognitive ability as measured by IQ tests has been an area of intensive study. The forerunner to this debate more than a century ago was Galton's (1969) classic work on "hereditary genius." Since then there has been a large number of studies, most of which have concentrated on g, or general cognitive ability. Table 1 summarizes the intraclass correlations for various categories of relatives. These data were originally taken from a review by Bouchard and McGue (1981) and have been updated by McGue et al. (1993) and Plomin et al. (1997). The data have to be interpreted against a background of the genetic and environmental influences shared by each category of relative pairs. Thus monozygotic (MZ) twins share all of their genes and, normally, share the environment in which there are reared. Dizygotic (DZ) twins, on the other hand, share only 50 % of their genes but it is assumed that, under normal circumstances, they share their environment of rearing to roughly the same extent as MZ twins (especially if we are concerned with same sex DZ pairs). Therefore, any greater similarity in MZ than in DZ pairs can be attributed to genetic influences. Thus, the MZ/DZ comparison in Table 1 immediately suggests a genetic effect. It is possible to go further than this in estimating the size of the genetic effect since it is easy to show

Table 1. Mean intraclass correlations for IQ[a]

MZ twins	0.86
DZ twins	0.6
MZ twins reared apart	0.78
sibs	0.47
sibs reared apart	0.24
adoptive sibs	0.25
adoptive sibs as adults	−0.01

[a] Data from McGue et al. 1993; Plomin et al. 1997. MZ, monozygotic; DZ, dizygotic.

that doubling the difference between the MZ and DZ correlations provides an approximate error estimate of heritability, or proportion of variants accounted for by genes (McGuffin et al. 1994; Plomin et al. 1997).

Of course, the "equal environment assumption" that enables this calculation to be made has been opened to criticism on the grounds that MZ twins may have more environmental sharing that their DZ counterparts. For example, in twin studies of various phenotypes, MZ pairs have been found more often than DZ pairs to share friends or be dressed alike as children or even to be placed in the same class at school (McGuffin et al. 1996). There are various ways of checking on the equal environment assumption (Plomin et al. 1997), but one of the most direct is to study MZ twins who have been separated early in life and who have been reared apart. Such pairs are rather rare, but Table 1 gives a correlation based upon two studies published within the past decade and containing a total of 93 pairs. As can be seen, the correlation is only a little less than for MZ twins reared together. Furthermore, as genes are the only source of resemblance for twins reared apart, the intraclass correlation can be interpreted as a direct measure of heritability. Therefore, the twins reared apart data give a rather higher estimate than those for the twins reared together (which from Table 1 would be $2 \times 0.26 = 0.52$).

Siblings, like DZ twins, share 50% of their genes on average but, it could be argued, tend to show less environmental sharing, being of different ages. The observed correlation for siblings in Table 1 is indeed slightly lower than that for DZ twins. Another suggestion that shared environment plays a role comes from the observation that siblings reared apart because one or both have been adopted away show a lower correlation than siblings reared together and, during childhood, adopted siblings show approximately the same correlation as genetic siblings who have been separated. However, it is of interest that this effect seems to pertain only in childhood, since adoptive siblings tested for IQ as adults show a correlation that is close to zero (McGue 1993).

This change over time is in keeping with twin data on twin pairs of different ages, which enable a developmental prospective to be taken. These data (Fig. 1) provide a somewhat counterintuitive pattern. One might expect that genetic effects would be at their greatest early in life, with environmental influences accumulating over time, causing the heritability of measured IQ to gradually diminish. Instead, what we see that there is a modest increase in the MZ correla-

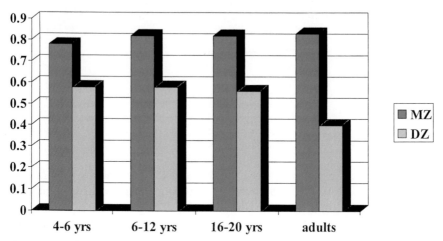

Fig. 1. IQ correlations in twins over time (data from McGue et al. 1993). MZ, monozygotic; DZ, dizygotic.

tion from early childhood to adult life, together with a marked decrease in the DZ correlation. In other words, the twin and adoptive data both suggest that the effects of environment that are shared within the family are moderately important in childhood but not in adult life. This finding has been most recently supported by a study of twins in late life, where cognitive ability was found to be substantially heritable, with no contribution from the shared environment (McClearn et al. 1997).

Emotionality and Depression

Breeders of domestic animals such as dogs or horses have long been aware that whether an animal is placid or "highly strung" is closely related to the temperament of its forebears. This belief had been confirmed in a more systematic way by breeding studies, mainly using rats or mice, showing that it is possible to select for high or low emotionality as measured by standard behavioural experiments such as the "open field" test.

Furthermore, it is possible to produce increasing divergence of strains for emotionality by continued selection over as many as 30 generations (DeFries et al. 1978), suggesting that many genes are involved. Nevertheless, it is becoming feasible to map quantitative trait loci (QTLs involved in emotionality in mice; Flint et al. 1995). In addition, studies of "knock out" mice have shown a role for particular genes in other behaviours such as increased aggression (Saudou et al. 1994; Nelson et al. 1995).

Attempts to devise standardised measures of emotionality within the normal range in humans have ranged wider than have studies on the measurement of cognitive ability. Nevertheless, by far the most studied measures of emotionality from a genetic prospective have dealt with the dimension of neuroticism-stability, often abbreviated to N (Eysenck and Eysenck 1976). Average intraclass correlations for N in MZ twins are around 0.46 and around 0.2 in DZ twins (Loehlin 1992). A curious finding in some early studies of twins reared apart (Shields 1962) was that MZ pairs reared apart were actually slightly more similar for N than MZ pairs reared together. More recent studies suggest only a modest difference between twins reared together and those raised apart (Loehlin 1992). Taken together the data suggest heritability in the region of 40 % to 50 % and, interestingly, metric analyses of MZ and DZ twins reared together suggest little or no effect of shared environment which, of course, is in keeping with the data on twins reared apart.

Whereas measures of neuroticism had been designed as quantitative measures of traits, mood disorders are episodic and are best considered as present/absent traits. Liability/threshold models (Falconer 1965) provide the most useful way of conceptualising the familial transmission of complex disorders such as depression. It is assumed that a continuous variable called liability is normally distributed in the population (or can be transformed to normality), but only those individuals whose liability at some point exceeds a certain threshold can be classified as affected. In the case of depressive disorders, where the principle feature is low mood or lack of enjoyment, high N scores would seem to be a reasonable predictor of vulnerability but are by no means a perfect measure of liability, since scores are markedly raised in individuals who are in a current episode of depression. It has also been shown that N scores are comparatively low in those relatives of depressed probands who have never been depressed, and are intermediate in relatives who have been depressed in the past but who are currently recovered. N in itself is largely a trait measure, but, when it comes to depression, it is "contaminated" by state. We'll now turn to consider the genetics of depressive disorder as such.

Genes and Depression

There is little doubt that affective disorder is familial. We owe much of our modern classification of psychiatric disorders to Kraepelin, who remarked on the marked heredity component in what was then called manic depressive insanity (Kraepelin 1922). However, following the suggestion of Leonhard (1959), affective disorder is nowadays usually subdivided into bipolar disorder, in which there are episodes of mania and depression (or less commonly just mania), and unipolar disorder, where there are episodes only of depression. The main differences between the two types in terms of familial distribution are that most recent studies find higher rates of affective disorder in the relatives of bipolar probands and an excess of both unipolar and bipolar disorders. By contrast, the rates of affective disorder in the relatives of unipolar probands have tended to be lower and, compared to the general population, there is an excess only of unipolar disorder (McGuffin et al. 1994).

The older twin study evidence (Zerbin-Rudin 1979; Bertelsen et al. 1977) also suggests a somewhat stronger genetic influence on bipolar disorder than on unipolar disorder. Hence, most studies attempting to uncover the molecular genetic bases of affective disorders have focused on bipolar disorder. Unfortunately, despite promising early results as long ago as 30 years using colour blindness as an excellent marker (Reich et al. 1969) and an initial striking finding using DNA markers (Egeland et al. 1987), no definite linkages have been confirmed (Risch and Botstein 1996). Nearly all such attempts have started with the hypothesis that major genes are involved at least in some families and have preferentially selected pedigrees with multiply affected in two or more generations. This strategy has of course being successful in other common disorders, including early onset Alzheimer (Sandbrink et al. 1995) disease and familial breast cancer (Daniels 1995). Although there is still a possibility of major genes subforms in a minority of families (e.g., Moises et al. 1995), there is an increasing consensus that even those recently reported linkages that have been partially replicated (Owen and McGuffin 1997) are at best detecting susceptibility loci that confer a fairly small relative risk.

By contrast with bipolar disorder, the genetics of unipolar depression has been somewhat underresearched. Given its high prevalence (Ormel et al. 1994) and the huge economic burden that depression causes worldwide, exceeding disorders such as ischaemic heart disease (Murray and Lopez 1997), this may seem surprising. However, the lack of research probably results from a combination of the familial aggregation of unipolar depression being less striking than for bipolar disorder and by an understandable emphasis on psychosocial correlation, since there is a clear and a consistent relationship with adversity (Brown and Harris 1978). However, recently there has been an increase in interest in the genetics of unipolar depression, in part resulting from the findings of twin studies.

In a large population-based study of female twins in the United States, Kendler et al. found high heritability for major depression in the region of 70% once the problems of unreliability of diagnosis were corrected for. Taking a rather narrower prospective on diagnosis, McGuffin et al. (1996) studied a sample of twins obtained via a hospital base register and again found heritability to be in the region of 70%. Neither study found any evidence of shared environmental effects. Somewhat in contrast, a population-based twin study of depressive *symptoms* during childhood and adolescents found strong influences of shared environment in children aged 11 years and younger but, in older children, twin similarities were explained entirely by genetic effects. The heritability was close to that for adult depressive *disorder* at around 70% (Thapar and McGuffin 1994, 1996). Therefore, given its ubiquity and the evidence of stronger genetic influences than were once suspected, unipolar depressive disorder would also seem to be a phenotype that needs to be tackled using molecular genetic methods.

Alternatives to Classical Linkage

There is broad agreement that standard approaches to detecting linkage are only likely to be successful for those phenotypes where there is an underlying major gene. Indeed LOD scores analysis (Morton 1955) were originally devised to detect linkage and estimate recumbernation for regular Mendelian traits. Although it is now possible to accommodate irregularities such as incomplete penetrants (Ott 1991), the success of analyses depends on the mode of transmission being specified accurately. In particular, mis-specification of the mode of transmission is likely to result in false negatives (i.e., exclusion of linkage when linkage is in fact present; Clerget-Darpoux 1991). Consequently, much recent attention has been focused on allele sharing methods in attempts to map susceptibility loci contributing to common diseases. Such methods, for example, using affected sib pairs require new assumptions about the mode of transmission. For any given marker locus the expectation of sharing 0, 1 or 2 alleles is, respectively, one quarter, one half and one quarter. Significant departure from expectations in the direction of increased allele sharing in affected sib pairs suggests that the marker is linked with a susceptibility gene contributing to the disease. Systematic genome SCANs using closely spaced microsatellite markers have been successful in implicating several loci in common diseases such as insulin-dependent diabetes (Cordell and Todd 1995). It is also possible to carry out analyses with highly polymorphic markers, without parental information without much loss of power, providing that marker gene frequencies can be estimated (Holmans 1993).

Analyses of sib pairs is also quantitative traits (Haseman and Elston 1972) based on the notion that the degree of similarity between siblings for phenotypic scores depends upon the degree of allele sharing at a quantitative trait locus. This approach has been elaborated (Fulker and Cardon 1994) and as been used, for example, to locate a gene contributing to a reading disability on the short arm of chromosome 6 (Cardon et al. 1994). There is also evidence of a contribute free locus on chromosome 15P (Grigorenko et al. 1997). Alternative approaches for mapping QTLs include variants components analyses in pedigrees, taking into account other types of relative pairs in addition to siblings (Amos 1994). However, such methods also depend for their success on the QTLs being detected having a comparatively large effect, i.e., explaining at least 10 % of the total phenotypic variants. For QTLs where the affect size is smaller than this, the sample size requirement may be enormous, running into several thousands sibling pairs (Fulker and Cardon 1994). Similarly for present/absent phenotypes such as diseases, affective sib pair approaches are efficient at detecting moderately large effects, i.e., a locus contributing a relative risk of three of more, but the sample size requirements become very large for loci contributing a relative risk of two or less.

Because of the problems of detecting small effects using affected sib pairs or QTLs linkage approaches, there has been increasing interest in using allelic association studies to detect genes involved in complex traits. These standard approaches to allelic association simply compare allele frequencies in a sample of cases with a disorder and a sample of ethnically matched controls. The simple case-controlled design is easily extended to quantitative traits by selecting "cases" who have an extreme score on the quantitative measure and comparing them with controls who have near-average scores. The attraction of case control allelic association studies, in addition to the simplicity of their design, is that they have long been known to be capable of detecting genes of small effects. For example Edwards (1965) pointed out that the by then well-replicated association between duodenal ulcer and blood group O accounted for only 1 % of the variance in liability to develop the disorder. Similarly, McGuffin and Buckland (1991) showed that the proportion of variance accounted for by HLA associations with various diseases was on the order of 3 %, and even the strongest known association between HLA B27 and ankylosing spondylitis was only around 20 %. The drawback of allelic association studies is that association only occurs if the marker *itself* contributes to the trait or is so close to the trait locus that the relationship is undisturbed over many generations of recombination, i.e., there is *linkage disequilibrium* (LD). Association studies are therefore potentially much more powerful than linkage approaches, but whereas linkage can be detected over fairly large distances, on the order of 10 centimorgans or more, association studies are "short-sighted" (Plomin et al. 1994) in that LD is only likely to occur in most populations at distances of less than 1 centimorgan.

Candidate Genes

The most direct approach to detecting allelic association is to focus on polymorphisms in or very close to genes that encode for proteins that are thought to be involved in the biochemical bases of the trait. In the case of emotional functions and mood disorders, secrotonergic pathways in the brain are thought to play a part. One of the major reasons for this is that drugs that augment secrotonergic transmission elevate depressed mood and, in particular, one of the commonest classes of drugs now used in the treatment of depression is the so-called selective secrotonin reuptake inhibitors (SSRIs), which include fluoxetine (Prozac). Such drugs block the action of serotonin transporter and there has hence been considerable interest in the human secrotonin transporter gene (hSERT), which has a candidate susceptibility gene for affective disorder. The hSERT gene maps to chromosomes 17q and contains two common polymorphisms, a variable number tandem repeat (VNTR) in introne 2 and a deletion/insertion in the promoter that has been shown to influence transcription activity and hence is functionally significant (Rees et al. 1997).

An association has been reported between the hSERT promoter polymorphisims and neuroticisms (Lesch et al. 1996). The VNTR in relation to mood disorder was first studied by Ogilivie et al. (1996), who found an association between unipolar depression and the comparatively rare allele 9. Subsequently, associations have been reported between affective disorder and the commoner allele 12 of the VNTR (Collier et al. 1996b), as well as with the promoter polymorphisms (Collier et al. 1996a). The association between the VNTR polymorphisms and bipolar affective disorder has subsequently been replicated, and there is some support for the association between unipolar disorder and allele 9 in a subset of patients (Rees et al. 1997).

When it comes to cognitive ability, the range of possible candidates is wide. Therefore, in a first attempt to search for association, Plomin et al. examined a total of 100 markers consisting of 18 multi-allelic and 72 bioallelic markers. These were selected on the broad basis that they were at or near genes that are expressed in the brain. A comparison of allele distributions was made in three groups of subjects having high, middle and low scores on IQ tests, and positive results were followed up on a replication sample. There were some promising preliminary findings including loci on chromosome 6P and a mitochondrial polymorphisms. However, the latter failed to replicate on an independent sample (Petrill 1998). The next stage of this project has, therefore, been to move on to a systematic whole genome search.

Genome Search for LD and Cognitive Ability

At the beginning of the 1990s we suggested (McGuffin et al. 1992) that, once a dense human genetic linkage map had been developed with very closely spaced markers, a whole genome search for linkage disequilibrium with complex traits would become feasible even in outbred populations. Although the standard text-book view has been that such an approach is not feasible (Strachan and Read 1996), the climate of opinion is changing (Risch and Merikangas 1996). A major problem, as we have noted, is that linkage disequilibrium is only likely to be detectable over very short distances of a centimorgan or less. Assuming genome length averaged across the sexes genome length of 3700 centimorgans, this means that around 2000 markers would be required for a genome search, ensuring that no susceptibility locus or QTLs was more than a centimorgan away from a marker.

We recently carried out a pilot search on chromosome 6q. This chromosome was selected because it is likely to be among the first large chromosomes whose DNA sequence will be completely determined as part of the Human Genome Project. To increase power with a manageable number of subjects, we studied 52 children with extremely high g (IQ greater than 160) and compared them with 50 controls with average g (mean IQ 101; Chorney et al. 1998). A total of 37 markers were studied, of which one, insulinlike growth factor-2 receptor (IGF2R), reached significance when the most common allele was compared with all other alleles in the two groups. This finding was replicated in a second high g and average control comparison. There were findings in the same directions, albeit of only marginal significance, in further groups selected for high math ability and high verbal ability. Combining all the data using Cochran's method, was a p value of less than 0.00003, which withstands correction for multiply testing. These results are clearly worth pursuing, but it should be pointed out that we have estimated that the IGF2R association accounts for less than 2 % of the variance in g. We cannot claim we have discovered the "gene for" IQ! Furthermore, the grid of markers used in this study on chromosome 6q was comparatively widely spaced and far from ideal for a systematic search. It nevertheless involved a large amount of genotyping shared between two collaborating laboratories.

We have therefore been exploring approaches that enable very high genotyping. Currently, developing technology to detect single nucleotide polymorphisms on micro arrays, is promising, but we have been pursuing an alternative approach, which is to carry out the initial genome screen using DNA pooling. We have already noted that the very minimum number of evenly spaced markers to perform a genome-wide search is 2000. Thus, in our current study searching for QTLs involved in IQ and in which we have 200 unrelated subjects with high IQ and 100 with average IQ, we would need to perform at least 600,000 individual genotypings (Daniels et al. 1998b). However, using DNA pooling we can simply combine the DNA from all the subjects in the high group and all those in the middle IQ group, reducing the number of genotypings in the initial phase to a more manageable 4000.

In brief, the approach is an extension of the no routine high throughput semi-automated genotyping using a DNA sequencer where simple sequence repeat polymorphisms (SSRPs) are fluorescently labelled and the resultant data are analysed by computer. There are a number of difficulties using SSRPs resulting from "stutter" bands and differential amplification. However, these are in part overcome using a simple statistic that compares the allele image patterns (AIPs) in the two groups by calculating a difference in areas (Δ AIP) by overlaying the traces derived from the two samples (Daniels et al. 1998a).

We have applied this approach using 147 roughly evenly spaced chromosome 4 markers.

The strategy is to carry out the initial screen with a fairly liberal criterion for statistical significance and then to attempt to replicate the positive results on an independent sample by carrying out conventional individual genotyping. On chromosome 4, we have interesting preliminary results with three significant QTL associations that withstand replication on an independent sample (Fisher et al. 1999).

Conclusions

There is a substantial genetic contribution to general cognitive ability, g, and to individual differences in emotion within the normal range and clinical depression.

However, common variations in cognitive ability and emotionality result from the combined effects of genes and environment. Furthermore, in each case there are likely to be multiple genes involved. Fortunately, techniques are now becoming available which make it feasible to search for multiple genes of small effects in such traits, and interesting preliminary findings include associations with the hSERT gene for neuroticisms and affective disorders and with polymorphisms on chromosomes 6q and 4 in cognitive ability.

References

Amos CI (1994) "Robust variance-components approach for assessing genetic linkage in pedigrees." Am J Human Genet 54:535–543

Bertelsen A, Harvald B, Gauge M (1977) A Danish twin study of manic-depressive disorders. Brit J Psychiat 130:330–351

Bouchard TJ, McGue M (1981) Familial studies of intelligence: a review. Science 212:1055–1059

Brown GW, Harris TO (1978) Social origins of depression: a study of psychiatric disorder in women. Tavistock Publications, London

Chorney MJ, Chorney K, Seese N, Owen MJ, Daniels J, McGuffin P, Thompson LA, Detterman DK, Benbow C, Lubinski D, Eley T, Plomin R (1998) A quantitative trait locus (QTL) associated with cognitive ability in children. Psychol Sci 9:159–166

Clerget-Darpoux F (1991) The uses and abuses of linkage analysis in neuropsychiatric disorder. In: McGuffin P, Murray R (eds) The new genetics of mental illness. Oxford, Butterworth-Heinemann, pp 44–57

Collier DA, Arranz MJ, Sham P, Battersby S, Vallada H, Gill P, Aitchison KJ, Sodhi M, Il T, Roberts GW, Smith B, Morton D, Murray RM, Smith D, Kirov G (1996a) The serotonin transporter is a potential susceptibility factor for bipolar affective disorder. Neuroreport 7:1675–1679

Collier DA, Stober G, Li T, Heils A, Catalano M, Di Bella D, Arranz MJ, Murray RM, Vallada HP, Bengel D, Muller CR, Roberts GW, Smeraldi E, Kirov G, Sham P, Lesch KP (1996b) A novel functional polymorphism within the promoter of the serotonin transporter gene: possible role in susceptibility to affective disorder. Mol Psychiat 1:453–460

Cordell HJ, Todd JA (1995) Multifactorial inheritance in type 1 diabetes. Trends Genet 11:499–504

Daniels J, Holmans P, Williams N, Turic D, McGuffin P, Plomin R, Owen MJ (1998a) A simple method for analysing microsatellite allele image patterns generated from DNA pools and its application to allelic association studies. Am J Human Genet 62:1189–1197

Daniels J, McGuffin P, Owen MJ, Plomin R (1998b) Molecular genetic studies of cognitive ability. Human Biol 70:277–291

Daves K (1995) Breast cancer genes. Further enigmatic variations. Nature 378:762–763

DeFreis JC, Gervais MC, Thomas EA (1978) Response to 30 generations of selection for open-field activity in laboratory mice. Behav Genet 8:3–13

Edwards JH (1965) The meaning of the associations between blood groups and disease. Ann Human Genet 29:77–83

Egeland JA, Gerhard DS, Pauls DL, Sussex JN, Kidd KK (1987) Bipolar affective disorders linked to DNA markers on chromosome 11. Nature 325:783–787

Eysenck HJ, Eysenck SB (1976) Manual of the EPQ (Eysenck Personality Inventory). Educational and Industrial Testing Service, San Diego

Falconer DS (1965) The inheritance of liability to certain diseases, estimated from the incidence among relatives. Ann Human Genet 29:51–76

Fisher PJ, Turic D, Williams NM, McGuffin P, Asherson P, Ball D, Craig I, Eley T, Hill L, Chorney K, Chorney MJ, Benbow CP, Lubinski D, Plomin R, Owen MJ (1999) DNA pooling identifies QTLs for general cognitive ability in children on chromosome 4. Human Molec Genet 8:915–922

Flint J, Corley R, De Fries JC, Fulker DW, Gray JA, Miller S, Collins AC (1995) A simple genetic basis for a complex psychological trait in laboratory mice. Science 269:5229, 1432–1435

Fulker DW, Cardon LR (1994) A sib-pair approach to interval mapping of quantitative trait loci. Am J Human Genet 54:1092–1103

Grigorenko EL, Wood FB, Meyer MS, Hart LA, Speed WC, Shuster A, Pauls DL (1997) Susceptibility loci for distinct components for development dyslexia on chromosomes 6 and 15. Am J Human Genet 60:27–39

Haseman JK, Elston RC (1972) The investigation of linkage between a quantitative trait and a marker locus. Behav Genet 2:3–19

Holmans P (1993) Asymptotic properties of affected-sib-pair linkage analysis. Am J Human Genet 52:362–374

Kendler JS, Neale M, Kessler R, Heath A, Eaves L (1993) The lifetime history of major depression in women: reliability of diagnosis and heritability. Ann Gen Psychiat 50:863–870

Kraepelin E (1922) Manic depressive insanity and paranoia (translated by RM Barclay). Edinburgh: E & S Livingstone

Leonhard K (1959) Aufteilung der Engoden Psychosen. Berlin, Akademic Verlag.

Lesch KP, Balling U, Gross J, Strauss K, Wolozin BL, Murphy DL, Riederer P (1994) Organisation of the human serotonin transporter gene. J Neural Trans (Gen Sect) 95:157–162

Loehlin JC (1992) Genes and environment in personality development, Sage, Newbury Park, CA

McClearn GE, Johansson B, Berg S, Pedersen NL, Ahern F, Petrill SA, Plomin R (1997) Substantial genetic influence on cognitive abilities in twins 80 or more years old. Science 276:1560–1563

McGue M (1993) From proteins to cognitions: the behavioral genetics of alcoholism. In: Plomin R, McClearn GE (eds) Nature, nurture and psychology. Washington, DC, American Psychological Association, pp 245–268

McGue M, Bouchard JT, Iacono WG, Lykken DT (1993) Behavioral genetics of cognitive ability: a life-span perspective. In: Plomin R, McLearn GE (eds) Nature, nurture and psychology. American Psychological Association, Washington, 59–76

McGuffin P, Buckland P (1991) Major genes minor genes and molecular neurobiology of mental illness. A comment on "quantitative trait loci and psychopharmacology" by Plomin, McLearn and Gora-Maslak. J Psychopharmacol 5:18–22

McGuffin P, Asherson P, Owen M, Farmer A (1994) The strength of the genetic evidence – is there room for an environmental influence in the aetiology of schizophrenia? Brit J Psychiat 164:593–599

McGuffin P, Katz R, Watkins S, Rutherford J (1996a) A hospital-based twin register of the heritability of DSM-IV unipolar depression. Arch Gen Psychiat 53:129–136

Morton NE (1955) Sequential tests for the detection of linkage. Am J Human Genet 7:277–318

Moises HW, Yang L, Kristbjarnarson H, Wiese C, Byerley W, Macciardi F, Arolt V, Blackwood D, Liu X, Sjögren B, Aschauer HN, Hwu H-G, Jang K, Livesley WJ, Kennedy JL, Zoega T, Ivarsson O, Bui M-T, Yu M-H, Havsteen B, Commenges D, Weissenbach J, Schwinger E, Gottesman II, Pakstis AJ, Wetterberg L, Kidd KK, Helgason T (1995) An international two-stage genome-wide search for schizophrenia susceptibility genes. Nature Genet 11:321–324

Murray CJ, Lopez AD (1997) Global morbidity, disability and the contribution of risk factors: global burden of disease study. Lancet 349:1476–1442

Nelson RJ, Demas GE, Huang PL, Fishman MC, Dawson VL, Dawson TM, Snyder SH (1995) Behavioural abnormalities in male mice lacking neuronal nitric oxide synthase. Nature 378:383–386

Ogilivie AD, Battersby S, Bubb VJ, Fink G, Harmar AJ, Goodwin GM, Smith CA (1996) Polymorphism in serotonin transporter gene associated with susceptibility to major depression. Lancet 347:731–733

Ormel J, Von Korff M, Ustun B (1994) Common mental disorders and disabilities across cultures: results from the WHO collaborative study on psychological problems in general health care. JAMA 282:1741–1748

Ott J (1991) Analysis of human genetic linkage. Baltimore, Johns Hopkins University Press

Owen MJ, McGuffin P (1997) Genetics and psychiatry. Br J Psychiat 171:201–202

Plomin R, Owen MJ, McGuffin P (1994) The genetic basis of complex human behaviors. Science 264:1733–1739

Plomin R, McClearn GE, Smith DL, Skuder P, Vignetti S, Chorney MJ, Chorney K, Kasarda S, Thompson LA, Detterman DK, Petrill SA, Daniels J, Owen MJ, McGuffin P (1995) Allelic association between 100 DNA markers and high versus low IQ. Intelligence 21:31–48

Plomin R, DeFries JC, McClearn GE, Rutter M (1997) Behavioral genetics. Freeman, New York

Rees M, Norton N, Jones I, McCandless F, Scourfield J, Holmans P, Moorhead S, Feldman E, Sadler S, Cole T, Redman K, Farmer A, McGuffin P, Owen MJ, Craddock N (1997) Association studies of bipolar disorder at the human serotonin transporter gene (hSERT; 5HTT). Mol Psychiat 2:396–402

Reich T, Clayton PJ, Winocur G (1969) Family history studies V. the genetics of mania. Am J Psychiat 125:1358–1369

Risch N, Botstein D (1996) A manic depressive history. Nature Genet 12:351–353

Risch N, Merikangas K (1996) The future of genetic studies of complex human diseases. Science 273:1516–1517

Sandbrink R, Hartmann T, Masters CL, Beyreuther K (1996) Genes contributing to Alzheimer's disease. Mol Psychiat 1:27–40

Saudou F, Amara DA, Dierich A, LeMeur M, Ramboz S, Segu L, Buhot MC, Hen R (1994) Enhanced aggressive behavior in mice lacking 5-HT1B receptor. Science 265:1875–1878

Shields J (1962) Monozygotic twins brought up apart and brought up together. Oxford University Press, London

Strachan T, Read AP (1996) Human molecular genetics. BIOS, Oxford

Thapar A, McGuffin P (1994) A twin study of depressive symptoms in childhood. Brit J Psychiat 165:259–265

Thapar A, McGuffin P (1996) A twin study of antisocial and neurotic symptoms in childhood. Psychol Med 26:1111–1118

Zerbin-Rudin E (1979) Genetics of affective disorders. In: Shou M, Strömgren E (eds) Origin prevention and treatment of affective disorders. Academic Press, London, pp 185–197

Gene Therapy: Promises, Problems and Prospects

I. M. Verma, L. Naldini, T. Kafri, H. Miyoshi, M. Takahashi, U. Blömer, N. Somia, L. Wang, and F. H. Gage

Gene therapy is a novel form of molecular medicine which will have a major impact on human health in the coming century. Although the advent of recombinant DNA technology in modern medicine will allow fetal genetic screening and genetic counseling, the vast majority of those born with the disease are likely to be helped by gene therapy approaches. The scope and definition of gene therapy have expanded in the past few years. In addition to the possibility of correcting inherited genetic disorders like cystic fibrosis, hemophilia and familial hypercholesterolemia, gene therapy approaches are being used to combat acquired diseases, like cancer, AIDS, infectious diseases, Parkinson's disease, and Alzheimer's disease. We are not, at this time, contemplating germ line gene therapy, due to the complex technical and ethical issues involved. We are interested in pursuing somatic cell gene therapy, which is exclusively for the benefit for the individual and cannot be passed on to the succeeding generation. The minimum requirement for gene therapy is sustained production of the therapeutic gene product without any harmful side effects (Anderson 1998; Verma and Somia 1997; Crystal 1995; Mulligan 1993; Leiden 1995).

Conceptually, gene therapy involves identifying appropriate DNA sequences and cell types and then developing suitable ways in which to get enough of the DNA into these cells. In 1990 the first clinical trials for gene therapy for adenosine deaminase (ADA) deficiency were carried out on two young girls at the National Institutes of Health (Anderson 1998). Over 200 clinical trials involving more than 6,000 patients are currently underway worldwide, yet there is not a single outcome that we can point to as a success story. Clearly, there are many technical, logistical, and, in some cases conceptual hurdles that need to be overcome before gene therapy becomes a routine practice in medicine.

The delivery system

The Achilles heel of gene therapy is gene delivery. The goal is to deliver genes efficiently and obtain sustained expression. Currently, the available gene delivery approaches can be divided into two categories: 1) physico-chemical methods, and 2) biological vectors. Physico-chemical methods involve: DNA transfection with calcium phosphate, DEAE Dextran, cationic lipids, etc.; direct DNA injection with or without the aid of cationic lipids; electroporation; ballistic guns

V. Boulyjenkov, K. Berg, Y. Christen (Eds.)
Genes and Resistance to Diseases
© Springer-Verlag Berlin Heidelberg 2000

delivering gold-coated DNA particles; microinjection; liposomes and receptor-mediated gene transfer (Table 1 in Mulligan 1993). All these approaches have benefits, but it is fair to say these methods are not ideal for efficient gene delivery and for production of sustained amounts of therapeutic gene product. For instance, an average individual secretes 5 μg/ml of factor IX protein in the plasma and the present methodology based on physico-chemical methods is just not quite up to par. This is not to say that better methods are not being developed. There are still many technical hurdles like delivery to the nucleus, integration, stability of the DNA, etc., which need to be overcome (Mulligan 1993). A majority of researchers have therefore taken advantage of biological vectors. For any large scale introduction of a biological entity to cells, the most likely approach will employ the use of viral vectors. After all, viruses have evolved into a specific machinery to deliver DNA to cells. High titer viruses (up to 10^6–10^{12} virus particles/ml) can be obtained (depending on the type of the virus) and it is easy to infect large numbers of cells. Additionally, pluripotent stem cells, which are rare ($1/10^3$ to 10^4 cells), have a chance to be successfully transduced with high titer viruses. Although there are many viral vectors, for our discussion we will focus on four types of viral vectors and indicate their major limitations and advantages.

Retroviral Vectors

Retroviruses are a group of viruses whose RNA genome is converted to DNA in the infected cell. The genome comprises three genes termed *gag*, *pol* and *env* which are flanked by elements called long terminal repeats (LTRs). These are required for integration into the host genome, and they define the beginning and end of the viral genome. The LTRs also serve as enhancer-promoter sequences – that is, they control expression of the viral genes. The final elements of the genome, called the packaging sequence (ψ), allows the viral RNA to be distinguished from other RNAs in the cell (Verma 1990; Coffin 1990).

By manipulating the viral genome, viral genes can be replaced with transgenes, such as the gene for factor IX. Transcription of the transgene may be under the control of viral LTRs or, alternatively, enhancer-promoter elements can be engineered in with the transgene. The chimeric genome is then introduced into a packaging cell, which produces all of the viral proteins (such as the products of the *gag*, *pol* and *env* genes), but these have been separated from the LTRs and the packaging sequence. So only the chimeric viral genomes are assembled to generate a retroviral vector (Danos and Mulligan 1988; Mann et al. 1983; Miller and Rosman 1989). The culture medium in which these packaging cells have been grown is then applied to the target cells, resulting in transfer of the transgene. Typically, a million target cells on a culture dish can be infected with one milliliter of the viral soup.

A critical limitation of retroviral vectors is their inability to infect non-dividing cells (Miller et al. 1990; Roe et al. 1983; Lewis and Emerman 1994), such as those that make up muscle, brain, lung and liver tissue. So, when possible, the cells from the target tissue are removed, grown in vitro, and infected with the recombinant retroviral vector. The target cells producing the foreign protein are then transplanted back into the animal. This procedure has been termed ex vivo gene therapy and our group has used it to infect mouse primary fibroblasts or myoblasts (connective tissue and muscle precursors, respectively) with retroviral vectors producing the factor IX protein. But within five to seven days of transplanting the infected cells back into mice, expression of factor IX is shut off (Dai et al. 1992; St. Louis and Verma 1988; Palmer et al. 1991). This transcriptional shut-off has even been observed in mice lacking a functional immune system (nude mice), and it cannot be due to cell loss or gene deletions because the transplanted cells can be recovered and, upon reculturing in vitro, produce factor IX protein (St. Louis and Verma 1988; Scharfmann et al. 1991).

Lentiviral Vectors

Lentiviruses also belong to the retrovirus family, but they can infect both dividing and non-dividing cells (Naldini and Verma 1998; Weinberg et al. 1991; Lewis et al. 1992). The best-known lentivirus is the human immunodeficiency virus (HIV), which has been disabled and developed as a vector for in vivo gene delivery. Like the simple retroviruses, HIV has three *gag, pol* and *env* genes, but it also carries genes for six accessory proteins termed *tat, rev, vpr, vpu, nef* and *vif* (Field et al. 1996). Using the retrovirus vectors as a model, lentivirus vectors have been generated, with the transgene ensconced between the LTRs and a packaging sequence (Naldini et al. 1996; Poeschla et al. 1996; Reiser et al. 1996) (Fig. 1). Some of the accessory proteins can be eliminated without affecting production of the vector or efficiency of infection (Zufferey et al. 1997; Kafri et al. 1997).The *env* gene product would restrict HIV-based vectors to infecting only cells that express CD4$^+$ so, in the vectors, this gene is substituted with *env* sequences from other RNA viruses that have a broader infection spectrum (such as glycoprotein from the vesicular stomatitis virus; Burns et al. 1993). These vectors can be produced presently at concentrations of $> 10^9$ virus particles per ml (Yee et al. 1994).

When lentivirus vectors are injected into rodent brain, liver, muscle, eye or pancreatic-islet cells, they give sustained expression for nearly the entire adult life of the animal (Blömer et al. 1997; Miyoshi et al. 1997) (Fig. 2). Unlike the prototypical retroviral vectors, the expression is not subject to shut-off. Little is known about the possible immune problems associated with the lentiviral vectors, but injection of 10^7 infectious units does not elicit a cellular immune response at the site of injection (Kafri et al. 1997). Furthermore, there seems to be no potent antibody response. So, at present, lentiviral vectors seem to offer an excellent opportunity for in vivo gene delivery with sustained expression.

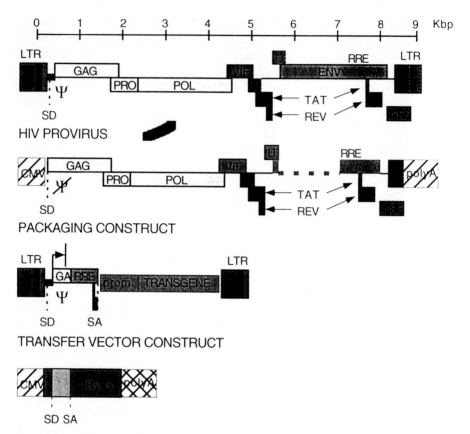

Fig. 1. Schematic drawing of the HIV provirus and the three constructs used to make lentiviral vector. The viral LTRs, the reading frames of the viral genes, the major 5' splice donor site (SD), the packaging sequence (ψ), and the Rev Response Element (RRE) are indicated. The packaging construct expresses the viral genes from heterologous transcriptional signals. Most of the envelope gene and the packaging sequence have been deleted. The transfer vector construct contains an expression cassette for the transgene flanked by the HIV LTRs. Downstream of the 5' LTR the vector contains the HIV leader sequence, the first 43 basepairs of the *gag* gene, the RRE element, and the splice acceptor sites of the third exon of *tat* and *rev*. A separate expression construct encodes a heterologous envelope to pseudotype the vector, here shown coding for the protein G of the Vesicular Stomatitis Virus (VSV.G). Only the relevant part of the constructs is shown. (Reprinted with permission from Naldini and Verma 1998).

Adenoviral Vectors

The adenoviruses are a family of DNA viruses that can infect both dividing and non-dividing cells, causing benign respiratory tract infections in humans (Strauss 1984; Chanrock et al. 1966; Berkner 1992). Their genomes contain over a dozen genes, and they do not usually integrate into the host DNA. Instead, they are replicated as episomal elements in the nucleus of the host cell (Shenk and Verify 1984; Kozarsky and Wilson 1993; Berkner 1988). Replication-deficient adenovirus vectors can be generated by replacing the *El* gene, which is essential for

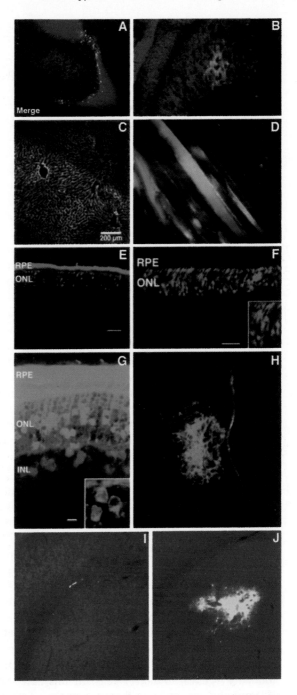

Fig. 2. In vivo gene (GFP) delivery with lentiviral vectors: *A, B,* rat brain; *C,* liver; *D,* muscle; *E, F,* retina; *G,* sin vector in retina and *H,* brain; *I,* rat striatum with tetracycline and *J,* 4 days after removal of tetracycline.

viral replication, with the gene of interest (for example, that for factor IX) and an enhancer-promoter element. The recombinant vectors are then replicated in cells that express the products of the *El* gene, and they can be generated in very high concentrations ($>10^{11}$–10^{12} adenovirus particles per ml; Dai et al. 1995).

Cells infected with recombinant adenovirus can express the therapeutic gene. Since essential genes for viral replication are deleted, the vector should not replicate. These vectors can infect cells in vivo, causing them to express very high levels of the transgene. Unfortunately, this expression lasts for only a short time (5–10 days post-infection). In contrast to the retroviral vectors, long-term expression can be achieved if the recombinant adenoviral vectors are introduced into nude mice or into mice that are given both the adenoviral vector and immunosuppressing agents (Dai et al. 1995; Yang et al. 1996). This finding indicates that the immune system is behind the short-term expression that is usually obtained from adenoviral vectors.

The immune reaction is potent, eliciting both the cell-killing "cellular" response and the antibody-producing "humoral" response. In the cellular response, virally infected cells are killed by cytotoxic T lymphocytes (Dai et al. 1995; Yang et al. 1996). The humoral response results in the generation of antibodies to adenoviral proteins, and it will prevent any subsequent infection if the animal is given a second injection of the recombinant adenovirus. Unfortunately for gene therapy, most of the human population will probably have antibodies to adenovirus from previous infection with the naturally occurring virus.

It is possible that the target cell contains factors that might trigger the synthesis of adenoviral proteins, leading to an immune response. To try to get around this problem, second-generation adenoviral vectors were developed, in which additional genes that are implicated in viral replication were deleted. These vectors showed longer-term expression, but a decline after 20–40 days was still apparent (Engelhardt et al. 1994). This idea has now been taken further with the generation of "gut-less" vectors; all of the viral genes are deleted, leaving only the elements that define the beginning and the end of the genome, and the viral packaging sequence. The transgenes carried by these viruses were expressed for 84 days (Chen et al. 1997). More recently, others have obtained even longer periods of expression.

There are considerable immunological problems to be overcome before adenoviral vectors can be used to deliver genes and produce sustained expression. The incoming adenoviral proteins that package DNA can be transported to the cytoplasm, where they are processed and presented on the cell surface, tagging the cell for destruction by cytotoxic T cells (Kafri et al. 1998). So adenoviral vectors are extremely useful if expression of the transgene is required for short periods of time. One promising approach is to deliver large numbers of adenoviral vectors containing genes for enzymes that can activate cell killing, or immunomodulatory genes, to cancer cells. In this case, the cellular immune response against the viral proteins will augment tumor killing. Finally, although immunosuppressive drugs can extend the duration of expression, our goal should be to manipulate the vector and not the patient.

Adeno-associated Viral Vectors

Adeno-associated virus (AAV) is a simple, non pathogenic, single-stranded DNA virus. Its two genes (*cap* and *rep*) are sandwiched between inverted terminal repeats that define the beginning and the end of the virus and contain the packaging sequence (Muzyczka 1992). The *cap* gene encodes viral capsid (coat) proteins, and the *rep* gene product is involved in viral replication and integration. AAV needs additional genes to replicate, and these are provided by a helper virus (usually adenovirus or herpes simplex virus).

The virus can infect a variety of cell types, and in the presence of the *rep* gene product, the viral DNA can integrate preferentially into human chromosome 19 (Kotin et al. 1991). To produce an AAV vector, the *rep* and *cap* genes are replaced with a transgene. Up to 10^{11}–10^{12} viral particles can be produced per ml, but only one in 100–1,000 particles is infectious (Samulski et al. 1989; Walsh et al. 1992; Chatterjee et al. 1992; Flotte et al. 1993; Flannery et al. 1997). Large-scale preparations of AAV vector are a bit laborious, due to the toxic nature of the *rep* gene product and some of the adenoviral helper proteins. We currently have no packaging cells in which all of the proteins can be stably provided. Vector preparations must also be carefully separated from any contaminating adenovirus, and AAV vectors can accommodate only 3.5–4.0 kilobases of foreign DNA. This will exclude larger genes like factor VIII. Recent advances in the use of deleted adeno plasmids as helper have made a big difference in making adeno-free AAV (Xiao et al. 1996). Finally, we need more information about the immunogenicity of the viral proteins, especially given that 80 % of the adult population have circulating antibodies to AAV. These considerations notwithstanding, AAV vectors containing human factor IX cDNA have been used to infect liver and muscle cells in immunocompetent mice. The mice produced therapeutic amounts of factor IX protein in their blood for over six months (Snyder et al. 1997; Herzog et al. 1997; Wang et al. 1999) (Fig. 3), confirming the promise of AAV as an in vivo gene therapy vector. AAV vectors have also been successfully used in transducing retinal cells (Flannery et al. 1997).

We have selected hemophilia (factor IX and VIII deficiency) as good model system because presently available vectors have the ability to successfully transduce both factor IX and factor VIII proteins in a variety of somatic tissues. The availability of both mouse and dog model systems make gene therapy studies of hemophilia a very attractive system to explore the problem of efficiency of transduction, duration of expression, immune problems and safety concerns. We firmly believe that gene therapy of hemophilia will serve as an important goal post towards the development of gene therapy as a commonly accepted practice in medicine.

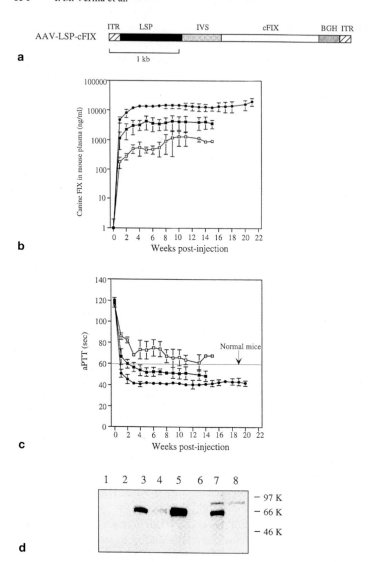

Fig. 3. Long-term expression of functional canine factor IX in hemophilia B mice. *a,* Schematic drawing of rAAV-LSP-cFIX vector. *b,* ELISA assays: Recombinant AAV-LSP-cFIX was delivered as a single intraportal injection into the liver of adult hemophilic C57BL/6 mice in a dose of 2×10^{11} (n = 4, □) or 5.6×10^{11} (n = 5, ●) vector genomes, or into the liver of adult hemophilic 129 mice at a dose of 2×10^{11} particles (n = 3, ■). Mice were bled weekly after vector administration. The canine factor IX concentration in the mouse plasma was determined by an ELISA assay. *c,* Functional factor IX activity in the mouse plasma was determined by an in vitro aPTT assay. The line in the middle shows the aPTT for normal mice. *d,* Western analysis. Plasma from C57BL/6 hemophilic mice 4 months postinjection with 2.8×10^{11} particles or 5.6×10^{11} particles of AAV-LSP-cFIX was analyzed by Western blot. Lanes: 1, naive mouse (0.1 µl of plasma); 2 (0.01 µl) and 3 (0.1 µl) plasma from mouse injected with 2.8×10^{11} particles; 4 (0.01 µl) and 5 (0.1 µl) of plasma from mouse injected with 5.6×10^{11} particles; 6 (0.01 µl) and 7 (0.1 µl) of normal canine plasma (Sigma): 8, 0.1 µl of plasma from a hemophilic dog (a gift from Dr. T. Nichols of the University of North Carolina, Chapel Hill). (Reprinted from Wang et al. 1999)

Acknowledgments

The authors are grateful to all their colleagues in the Verma and Gage labs for their sustained interest and support.

All the work presented here has been published previously. Furthermore, an identical manuscript has been submitted for publication in different conference proceedings.

This work is support by grants from the NIH, the March of Dimes Foundation for Birth Defects, and the H.N. and Frances C. Burger and The Wayne and Gladys Valley Foundations. Dr. Verma is an American Cancer Society Professor of Molecular Biology.

References

Anderson WF (1998) Human gene therapy. Nature 392 (Suppl), 25–30

Berkner KL (1988) Development of adenovirus vector for expression of heterologous genes. Biotechniques 6:616–629

Berkner KL (1992) Expression of heterologous sequences in adenoviral vectors In: Muzycka N (ed) Current topics in microbiology and immunology. Springer-Verlag, Berlin, 39–66

Blömer U, Naldini L, Kafri T, Trono D, Verma IM, Gage FH (1997) Highly efficient and sustained gene transfer in adult neurons with a lentiviral vector. J Virol 71:6641–6649

Burns JC, Friedman T, Driever W, Burrascano M, Yii J-K (1993) Vesicular stomatitis virus G protein pseudotyped retroviral vectors: concentration to very high titer and efficient gene transfer into mammalian and nonmammalian cells. Proc Natl Acad Sci USA 90:8033–8037

Chanrock RM, Ludwig W, Heubner RJ, Cate TR, Chui L-W (1966) Immunization with selective infection with type 4 adenovirus grown in human diploid tissue culture. I: safety and lack of oncogenicity and tests for potency in volunteers. JAMA 195:151–165

Chatterjee S, Johnson PR, Wong KK, Jr. (1992) Dual target inhibition of HIV-1 in vitro by means of an adeno-associated virus antisense vector. Science 258:1485–1488

Chen HH, Mack LM, Kelly R, Ontell M, Kochanek S, Clemens PR (1997) Persistence in muscle of an adenoviral vector that lacks all viral genes. Proc Natl Acad Sci USA 94:1645–1650

Coffin JM (1990) Retroviridae and their replication In: Fields BN Knipe DM et al. (eds) Virology. Second Edition, Raven Press, Ltd., New York, 1437–1500

Crystal RG (1995) Transfer of genes to humans: early lessons and obstacles to success. Science 270:404–410

Dai Y, Roman M, Naviaux RK, Verma IM (1992) Gene therapy via primary myoblasts: long-term expression of factor IX protein following transplantation in vivo. Proc Natl Acad Sci USA 89:10892–10895

Dai Y, Schwartz EM, Gu D, Zhang WW, Sarveknick N, Verma IM (1995) Cellular and humoral immune responses to adenoviral vectors containing factor IX gene: Tolerization factor IX and vector antigens allows for long-term expression. Proc Natl Acad Sci USA 92:1401–1405

Danos O, Mulligan RC (1988) Safe and efficient generation of recombinant retroviruses with amphotropic and ecotropic host ranges. Proc Natl Acad Sci (USA) 85:6460–6464

Engelhardt JF, Ye X, Doranz B, Wilson JM (1994) Ablation of E2A in recombinant adenoviruses improves transgene persistence and decreases inflammatory response in mouse liver. Proc Natl Acad Sci USA 93:6196–6200

Field BN, Knipe DM, Howley PM (eds) (1996) Virology. Lippincott-Raven, Philadelphia PA

Flannery JG, Zolotukhin S, Vaquero M, La Vail MM, Muzyczka N, HAuswirth WW (1997) Efficient photoreceptor-targeted gene expression in vivo by recombinant adeno-associated virus. Proc Natl Acad Sci USA 94:6916–6921

Flotte TR, Afione SA, Conrad C, McGrath SA, Solow R, Oka H, Zeitlin PL, Guggino WB, Carrer BJ (1993) Stable in vivo expression of the cystic fibrosis transmembrane conductance regulator with an adeno-associated viral vector. Proc Natl Acad Sci USA 90:10613–10617

Herzog RW, Hagstrom JN, Kung SH, Tai SJ, Wilson JM, Fisher KT, High KA (1997) Stable gene transfer and expression of human blood coagulation factor IX after intramuscular injection of recombinant adeno-associated virus. Proc Natl Acad Sci USA 94:5804–5809

Kafri T, Blömer U, Gage FH, Verma IM (1997) Sustained expression of genes delivered directly in liver and muscle by lentiviral vectors. Nat Gen 17:314–317

Kafri T, Morgan D, Krahl T, Sarrenirck N, Sherman L, Verma IM (1998) Cellular immune response to adenoviral vector infected cells does not require *de novo* viral gene expression: implications for gene therapy. Proc Natl Acad Sci USA 95:11377–11382

Kotin RM, Menninger JC, Ward DC, Berns KI (1991) Mapping and direct visualization of a region-specific viral DNA integration site on chromosome 19q13-qtr. Genomics 10:81–834

Kozarsky FK, Wilson JM (1993) Gene therapy: adenovirus vectors. In: Current opinions in genetics and development. 499–503

Leiden JM (1995) Gene therapy-promises, pitfalls and prognosis. New Engl J Med 333:871–873

Lewis PF, Emerman M (1994) Passage through mitosis is required for oncoretroviruses but not for the human immunodeficiency virus. J Virol 68:510–516

Lewis P, Hensel M, Emerman M (1992) Human immunodeficiency virus infection of cells arrested in the cell cycle. EMBO J 11:3053–3058

Mann R Mulligan RC, Baltimore D (1983) Construction of a retrovirus packaging mutant and its use to produce helper-free defective retroviruses. Cell 33:153–159

Miller AD, Rosman GJ (1989) Improved retroviral vectors for gene transfer and expression. Biotechniques 7:980–982, 984–986, 989–990

Miller DG, Adam MA, Miller AD (1990) Gene transfer by retrovirus vector occurs only in cells that are actively replicating at the time of infection. Mol Cell Biol 10:4239–4242

Miyoshi H, Takahashi M, Gage FH, Verma IM (1997) Stable and efficient gene transfer into the retina using a lentiviral vector. Proc Natl Acad Sci USA 94:10319–10323

Mulligan RC (1993) The basic science of gene therapy. Science 260:926–932

Muzyczka N (1992) Use of adeno-associated virus as a general transduction vector for mammalian cells In: Current topics in microbiology and immunology. Springer-Verlag, Berlin, 97–123

Naldini L, Verma IM (1998) In: (Friedman T, ed) The development of human gene therapy. CSHL Press, Cold Spring Harbor, 47–60

Naldini L, Blomer U, Gallay P, Ory D, Mulligan P, Gage FH, Verma IM, Trono D (1996) In vivo gene delivery and stable transduction of nondividing cells by a lentiviral vector. Science 272:263–267

Palmer TD, Rosman GJ, Osborne WRA, Miller AD (1991) Genetically modified skin fibroblasts persist long after transplantation but gradually inactivate introduced genes. Proc Natl Acad Sci USA 88:1330–1334

Poeschla E, Corbeau P, Wong-Staal F (1996) Development of HIV vectors for anti-HIV gene therapy. Proc Natl Acad Sci USA 93:11395–11399

Reiser J, Harmison G, Kluepfel-Stahl S, Brady RO, Karlsson S, Schubert M (1996) Transduction of non-dividing cells using pseudotyped defective high-titer HIV type 1 particles. Proc Natl Acad Sci USA 93:15266–15271

Roe T, Reynolds TC, Yu G, Brown PO (1993) Integration of murine leukemia virus DNA depends on mitosis. EMBO J 12:2099–2108

Samulski RJ, Chang L-S, Shenk T (1989) Helper-free stocks of recombinant adeno-associated viruses: normal integration does not require viral gene expression. J Virol 63:3822–3828

Scharfmann R, Axelrod JH, Verma IM (1991) Long-term in vivo expression of retrovirus-mediated gene transfer in mouse fibroblast implants. Proc Natl Acad Sci USA 88:4626–4630

Shenk TJ, Williams (1984) Genetic analysis of adenoviruses. In: Current topics in microbiology and immunology. Springer-Verlag, Berlin, 1–39

Snyder RO, Miao CH, Parijn GA, Spratr SK, Danos O, Nagy D, Gown AM, Winkler B, Meuse L, Cohen LK, Thompsen AR, Kay MA (1997) Persistent and therapeutic concentrations of human factor IX in mice after hepatic gene transfer of recombinant AAV vectors. Nat Genet 16:270–276

St. Louis D, Verma IM (1988) An alternative approach to somatic cell gene therapy. Proc Natl Acad Sci USA 85:3150–3154

Strauss SE (1984) In: Ginsberg HS (ed) The adenoviruses. Plenum Press, New York, 451–496

Verma IM (1990) Gene therapy. Sci Amer 262:68–84

Verma IM, Somia N (1999) Gene therapy: promises, problems and prospects. Nature 389:239–242

Walsh CE, Liu JM, Xiao X, Young NS, Nienhuis AW (1992) Regulated high level expression of a human γ-globin gene introduced into erythroid cells by adeno-associated virus vector. Proc Natl Acad Sci USA 89:7257–7261

Wang L, Takabe K, Bidlingmaier SM, Ill CR, Verma IM (1999) Sustained correction of bleeding disorder in hemophilia B mice by gene therapy. Proc Natl Acad Sci USA 96:3906–3910

Weinberg JB, Matthews TJ, Cullen BR, Malim MH (1991) Productive human immunodeficiency virus type 1 (HIV-1) infection of nonproliferating human monocytes. J Exper Med 174:1477–1382

Xiao X, Li J, Samulski RJ (1996) Efficient long-term gene transfer into muscle tissue of immunocompetent mice by adeno-associated virus vector. J Virol 70:8098–8108

Yang Y, Greenough K, Wilson JM (1996) Transient immune blockade prevents formation of neutralizing antibody to recombinant adenovirus and allows repeated gene transfer to mouse liver. Gene Ther 3:412–420

Yee J-K, Miyanohara A, Lalorte P, Bovic K, Burns JC, Friedmann T (1994) A general method for the generation of high-titer, pantropic retroviral vectors: highly efficient infection of primary hepatocytes. Proc Natl Acad Sci USA 91:9564–9568

Zufferey R, Nagy D, Mandel RJ, Naldini L, Trono D (1997) Multiply attenuated lentiviral vectors achieves efficient gene delivery in vivo. Nat Biotechnol 15:871–875

"Good Gene"/"Bad Gene"

V. I. Ivanov

> "You may freely eat of every tree of the garden;
> but of the tree of the knowledge of good and evil
> you shall not eat, for in the day that you eat of it
> you shall die".
> Genesis I, 16–17.

Summary

The manifestation of many genes and their alleles shows broad variation depending on genetic, intraorganismal and environmental background. The resistance versus susceptibility and the predisposition to a disease are not exceptions to the rule. Quite a number of cases have been thoroughly elaborated in which the level of an organisms susceptibility to a disease was found to be dependent on certain genes and genotypes, including many polymorphic systems. In some cases the physiological, cellular, genetic and molecular mechanisms of gene interaction in morbid trait development have already been revealed. Recent examples illustrate the applicability of the general features of variation in hereditary trait manifestation to human diseases and show that the genes may be "good" or "bad" depending on the actual conditions of an organisms development.

Introduction

Distinguishing between "good" and "bad" is one of the most ancient problems, which people have never ceased posing and trying to solve when studying any natural or artificial phenomenon. This struggle applies to the notion of genes, as well, since it was introduced by W. Johannsen in 1909. At that time, a rather limited number of examples of gene expression was available in a few plant and animal species and not more than a dozen similar examples were found in humans. But even then it was clear that some genes and some of their allelomorphs (alleles) promote the development of "good," i.e., favourable traits and characters while others cause rather "bad" effects. It was soon discovered that the manifestation of many genes and their alleles shows broad variation depending on genetic, intraorganismal and environmental background. The resistance versus susceptibility and the predisposition to a disease are not exceptions to the rule. Moreover, quite a number of cases have since been thoroughly elaborated in which the level of an organism's susceptibility to a disease was found to be dependent on certain genes and genotypes, including many polymorphic systems. In some cases the physiological, cellular, genetic and molecular mechanisms of gene interaction in morbid trait development have already been revealed. Thus, whether a gene is "good" or "bad" is determined not only by its

V. Boulyjenkov, K. Berg, Y. Christen (Eds.)
Genes and Resistance to Diseases
© Springer-Verlag Berlin Heidelberg 2000

intrinsic, invariant characteristics but also by a particular combination of such parameters as the general genetic constitution of the organism in question and its intraorganismal and environmental conditions.

Basic Features of Variation of Hereditary Traits

At the border of the 19th and the 20th centuries, the Mendelian rules of 1) uniformity of the first generation garden pea (*Pisum sativum*) hybrids, 2) segregation of traits in succeeding generations and backcrosses of the plants; and 3) random combination of different traits in multihybrid crosses (Mendel 1866) were independently and nearly simultaneously rediscovered by several authors in a number of plant species (Correns 1900; von Tschermak 1900; de Vries 1900), in poultry (Bateson and Sounders 1902) and in mice (Cuenot 1902). A year later Sutton (1903) inferred that chromosome behaviour in somatic and germ line cell divisions (mitosis and meiosis, respectively) constitutes the physical, cytological basis of Mendelian inheritance.

These findings induced a dense flow of hybridization experiments in diverse plant and animal species in which segregation of many qualitative and quantitative characters was studied. Some of these hybridization experiments showed certain deviations from Mendelian rules, such as linked versus random recombination of some sets of traits (Bateson et al. 1905; Morgan et al. 1915), modification of expression (enhancement and/or suppression) of some genes by others (Bridges 1919), multiple (pleiotropic) manifestations of certain genes (Plate 1910) and heterogeneous determination of others (Timofeeff-Ressovsky 1934), etc. An example of pleiotropy versus heterogeneity in the action of some human morbid genes is presented in Figure 1. In at list five distinct and genetically different conditions, some features of Marfanoid habitus are observed; any of these conditions shows a number of morbid traits, like five major components of Marfan syndrome.

In 1925 Timofeeff-Ressovsky introduced two quantitative indices measuring the variation of phenotypic expression of genotypes: 1) **penetrance,** indicating the fraction of individuals of respective genotype showing any signs of that genotype in their appearance, and 2) **expressivity,** reflecting the range of phenotypic variation of the genotype expression in penetrant individuals (Timofeeff-Ressovsky 1925). Very soon the terms penetrance and expressivity were applied to the description and analysis of morbid trait variation in human beings (Timofeeff-Ressovsky and Vogt 1926).

Further elaborating the theory of variation of genotype expression, Timofeeff-Ressovsky suggested a general scheme (Fig. 2) according to which the particular phenotype is a result of a wholistic complex process in which the variable that is in itself the primary action of the gene in question is further modified depending on the actual parameters of genomic, intraorganismal and environmental media (Timofeeff-Ressovsky 1940).

Along the same lines, Dobzhansky (1976) discussed the relativity of describing human traits as being genetic or environmental, normal or morbid: "The

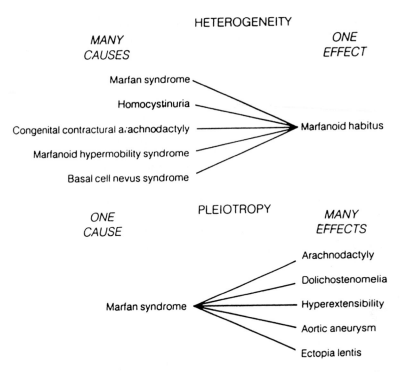

HETEROGENEITY

MANY CAUSES   *ONE EFFECT*

Marfan syndrome

Homocystinuria

Congenital contractural arachnodactyly   Marfanoid habitus

Marfanoid hypermobility syndrome

Basal cell nevus syndrome

ONE CAUSE   **PLEIOTROPY**   *MANY EFFECTS*

        Arachnodactyly

        Dolichostenomelia

Marfan syndrome   Hyperextensibility

        Aortic aneurysm

        Ectopia lentis

Fig. 1. Heterogeneity and pleiotropy in manifestation of hereditary traits. (after Cohen 1982)

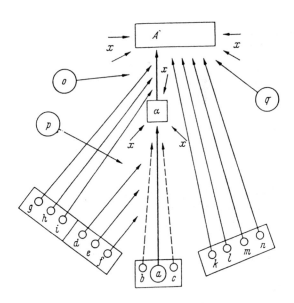

Fig. 2. General scheme of gene manifestation: *a*, major gene primarily determining the definitive trait *A* through some intermediate products *α*; *b–c, d–f, g–i, k–n*, groups of genes participating in phenotypic expression of *α*; *x*, unspecified factors of intraorganismal medium causing variation in the realization of gene *a* in trait *A* at different steps of the process; *o, p, q*, environmental agencies. Rectangle in the top symbolizes the width of the field (reaction norm) within which the definitive trait *A* may vary. (after Timofeeff-Ressovsky 1940)

same disease may be genetic or environmental. Consider the following hypothetical situation in which every human being carried the genes for phenylketonuria or for diabetes. Under such conditions, a nutrition nearly free of phenylalanine, or public health program supplying insulin, would then be the "normal" environment. The diseases would appear only if by some accident phenylalanine got in the diet or if insulin were in short supply. But then phenylketonuria and diabetes would no longer be genetic, they would be environmental diseases! At the same time, for persons free of phenilketonuric or diabetic genes, phenylalanine-free diets and insulin injection are uncalled for. They might be even harmful".

An interesting example of morbid phenotype variation in man was presented by Kozlova and Prytkov (1986), who found among Azerbaijan mountaineers a large, six-generation kindred of 399 persons including 92 affected with autosomal dominant Ehlers-Danlos syndrome (Fig. 3). In spite of all the members of the kindred being descendants of one common ancestor, three clinical forms of the syndrome were found intermingled in generations among them (Table 1).

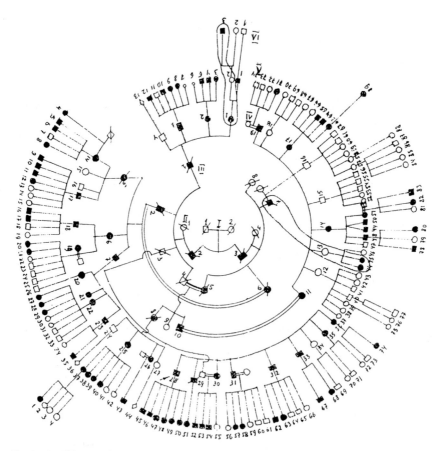

Fig. 3. An Ehlers-Danlos pedigree. (after Kozlova and Prytkov 1986)

Table 1. Occurrence of Ehlers-Danlos syndrome clinical forms (I, II and III) in parents and children.

Children	Parents		
	I	II	III
I	30	7	–
II	12	9	–
III	8	3	–

Resistance and Susceptibility to Diseases in Humans

In this section some recent examples are presented which illustrate the applicability of the above-mentioned general features of variation in manifestation of hereditary traits to human diseases (Table 2).

Genes shown to affect risk factors or protective factors with respect to coronary heart disease (CHD) have been identified at the APOB, APOAI, LPA, LDLR, APOE and CETP loci. Rare mutations may have a major effect, whereas genes belonging to normal polymorphism have only a moderate effect. Even genes with only a slight effect can be clinically important in combination with other genes or life-style factors. There is gene-to-gene interaction between LDLR and APOE genes. Important risk factors determined by genes as well as by environmental factors are homocystein and fibrinogen (Berg 1998).

In the search for the interaction between insertion/deletion (I/D) angiotensin I-converting enzyme (ACE) and M235T angiotensinogen (AGT) gene polymorphic alleles in causing essential hypertension, 365 Caucasians from the Czech Republic (202 normotensives and 163 hypertensives) were examined. No association of single gene allelic variants with essential hypertension was found in the population. However, the association of the DDMM genotype with essential hypertension was proven. To the contrary, IITT was found more frequently in normotensive subjects. The interaction of the I/D ACE and M235T AGT polymorphic alleles can contribute to essential hypertension (Vasku et al. 1998).

In other studies on CHD summarized by V. A. Spitsyn (personal communication), serum markers Le (a-b-) of the Lewis system, Gm (1-), PI M1M1 of the proteinase inhibitor system and Se of the ABH secretion system were found to be risk factors, whereas Le(a-b+), Gm (1+), PI M1M2 and se (ABH) showed protective effects. Also, DNA markers M+ and M+M+ of the Msp1 polymorphic system were found more frequently in CHD patients than in the control group. On APO examination APO B 30 and APO CII 30 were less frequently observed in CHD patients. On the contrary, APO B 32 may be considered a risk factor. APO BL alleles were found to be unfavourable only in combination with APO B 36, but not APO B 34. Combination of unfavourable genotypes APO B 36L and APO B LL with a favourable APO CII 30 decreased the risk. Insertion-deletion polymorphism in angiotensin-converting enzyme is also relevant in CHD development: allele D and genotype DD are distinct risk factors.

Carcinogenic heterocyclic amines are activated by N-acetyltransferase (NAT) enzymes, encoded by NAT1 and NAT2, to genotoxic compounds that can form

Table 2. Protective (P) and/or risk (R) genetic (g) and environmental (e) factors and their interactions (g-g, g-e) in some diseases.

Diseases	P		R			
	g	g-g	g	e	g-g	g-e
CHD	ABH se ACE D, ACE DD		ABH Se			
		APO 36– APO CII 30, APO B LL- APO C II 30	APO B 32		APO BL- APO B 36	
	GM 1+ Le a-b+		GM 1- Le a-b- Msp1 M+ Msp1 M+M+			
	PI M1M2		PI M1M1			
EH		ACE TT- AGT II			ACE MM- AGT DD	
Breast cancer			1q-, 3p-, 11 p-, 13p-, 16q-, 17p-, 18q-			
Colorectal cancer						NAT rapid- red meat
Cancers			17p-			
Asbestosis	C3 FS HP*2		C3FF HP*1			
Fluorosis	ACP1*BB Cerumen*W PGM1+		ACP1*A Cerumen*d PGM1-			
Silicosis	Gm1- HLA*A3, B12, B35		HLA*A28, B18 PGM1 2–2			
	PI M1M1					
Narcolepsy-cataplexy			HLA DQB1	25–31% MZ		
Osteoporosis		PvuII ER (–/–) – VDR bb			PvuII ER (–/–) – VDR BB	

DNA adducts in the colon epithelium. The relation of polymorphisms in the genes coding for both enzymes to a risk of colorectal cancer and the gene-environment interaction with red meat intake were examined among participants in a prospective study. Baseline blood samples from 212 men subsequently diagnosed with colorectal cancer during 13 years of follow-up were genotyped, along with 221 controls. NAT genotypes were analyzed by a polymerase chain reaction (PCR)-restriction fragment length polymorphism (RFLP) method. There was no overall independent association of NAT acetylation genotypes and colorectal cancer risk. The relative risks for the rapid acetylation genotype were 0.93 for NAT1, 0.80 for NAT2, and 0.81 for NAT1/NAT2. A stronger association of red meat intake with cancer risk was observed among NAT rapid acetylators, especially among men 60 years old or older. These prospective data suggest that polymorphisms in the NAT genes confer differential susceptibility to the effect of red meat consumption on colorectal cancer risk (Chen et al. 1998).

The loss of heterozygosity due to chromosome deletions is often observed in various cancers. Thus, 17p- was found in tumors affecting colon, breast, brain, and bladder; deletions in 1q, 3p, 11p, 13p, 16q, 17p, or 18q arms are frequent in breast cancer, indicating some involvement of genes located in these arms in breast carcinogenesis. The same is true for genes located in 3p, 13q or 17p in persons affected with lung cancer.

Numerous associations between genetic markers and occupational diseases were found by V. A. Spitsyn and his colleagues of the Research Centre for Medical Genetics, Moscow, Russia. In studies on asbestosis they found a protective effect of 3rd complement component C3 FS allele and haptoglobin HP 2 versus risk uprising action of C3 FF and HP 1. In a similar study of fluorosis in aluminium shop workers' blood, erythrocyte enzyme variants ACP1 BBB and PGM1+ and cerumen W were found more frequently in healthy persons, whereas ACP1 A, PGM1- and cerumen d were more specific for those affected by fluorosis. Associations were also found between predisposition to silicosis and Gm1-, HLA A3, B12, B35 and PI M1M1 genotypes whereas HLA A28 and B18 as well as PGM1 2-2 carriers were more resistant to the disease.

Genetic and environmental risks were found in narcolepsy-cataplexy, a disabling sleep disorder whose development involves environmental factors acting on a specific genetic background. The importance of environmental factors is evidenced by only 25 to 31% of monozygotic twins are concordant for narcolepsy. One of the predisposing genetic factors is located in the major histocompatibility complex (MHC) DQ region. More than 85% of all narcoleptic patients with definite cataplexy share a specific human leucocyte antigen (HLA) allele, HLA DQB1 0602, compared with 12 to 38% of the general population, as evaluated in various ethnic groups. Genetic factors other than HLA are also likely to be involved. Even if genuine multiplex families are rare, 1 to 2% of the first-degree relatives of narcolepsy patients manifest the disorder, compared with 0.02 to 0.18% in the general population (Mignot 1998).

Low bone mineral density (BMD) is a major risk factor for development of osteoporosis. Increasing evidence suggests that attainment and maintenance of peak bone mass as well as bone turnover and bone loss have strong genetic determinants. The association between BMD levels and their change over a three-year period and polymorphisms of the estrogen receptor (ER), vitamin D receptor (VDR), type I collagen, osteonectin, osteopontin, and osteocalcin genes was examined in pre- and perimenopausal women of the Michigan population-based longitudinal study of BMD. Body composition measurements, reproductive hormone profiles, bone-related serum protein measurements, and life-style characteristics were also available on each woman. Based on an evaluation of the women, ER genotypes were significantly predictive of both lumbar spine and total body BMD levels, but not their change over the three-year period examined. The VDR BsmI RFLP was not associated with baseline BMD, change in BMD over time, or any of the bone-related serum and body composition measurements in the 372 women in whom it was evaluated. Likewise, none of the other polymorphic markers was associated with BMD measurements. However, a significant gene-to-gene interaction effect for the VDR locus and PvuII and XbaI polymorphisms was identified, which impacted BMD levels. Women who had the (–/–) PvuII ER and bb VDR genotype combination had a very high average BMD, whereas individuals with the (–/–) PvuII ER and BB VDR genotype had significantly lower BMD levels. This contrast was not explained by differences in serum levels of osteocalcin, parathyroid hormone, 1,25-dihydroxyvitamin D, or 25-dihydroxyvitamin D. These data suggest that genetic variation at the ER locus, singly and in relation to the vitamin D receptor gene, influences attainment and maintenance of peak bone mass in younger women, which in turn may render some individuals more susceptible to osteoporosis than others (Willing et al. 1998).

Animal studies have suggested an important role for the homeobox-containing gene MSX1 in limb, oralfacial, and cardiac malformations. A study of 516 Caucasians with isolated birth defects registered in the Maryland Birth Defects Reporting and Information System (BDRIS) reported an association between a dinucleotide repeat polymorphism in MSX1 and isolated limb deficiency. Frequencies of rare alleles at the MSX1 locus were found to be significantly higher among 34 infants with limb deficiency compared to 482 infants with other isolated birth defects (oral clefts, dislocation of hip, clubfoot, hypospadias, polydactyly, or syndactyly). Infants carrying the rare alleles had an almost five-fold higher risk of a limb deficiency when the mother reported smoking during pregnancy, compared to infants who were homozygous for the common allele and whose mothers did not smoke. The significance of this apparent gene-environment interaction is attributed to infants with malformation of the lower limb (Hwang et al. 1998).

Conclusion

Summarizing the series of examples of variation in resistance and/or susceptibility to various genetic and non-genetic diseases, it may be suggested that referring to certain genes as being "good" or "bad" per se is a rather primitive way of thinking. Favourable or hazardous effects of genes depend to a great extent not only upon which alleles of the gene were inherited by a person from his or her parents but also upon the actual circumstances in which the long distance was passed from the gene to a definitive normal or morbid character, as well as what were the intragenomic, intracellular, intraorganismal and environmental conditions during the course of the organisms development, from preconception to the moment in question.

References

Bateson W, Saunders ER (1902) Experimental studies in the physiology of heredity. Rept Evol Comm Royal Soc I

Bateson W, Saunders ER, Punnett RC (1905) Experimental studies in the physiology of heredity. Rept Evol Comm Royal Soc II 1–131

Berg K (1998) [Molecular biology in the diagnosis of cardiovascular diseases] Molekylaerbiologisk diagnostikk ved hjerte- og karsykdommer. Tidsskr-Nor-Laegeforen 118:2370–2374

Bridges CB (1919) Specific modifiers of eosin eye color in *Drosophila melanogaster*. J Exp Zool 28:337–384

Chen J, Stampfer MJ, Hough HL, Garcia-Closas M, Willett WC, Hennekens CH, Kelsey KT, Hunter DJ (1998) A prospective study of N-acetyltransferase genotype, red meat intake, and risk of colorectal cancer. Cancer Res 58:3307–3311

Cohen MM (1982) The child with multiple birth defects. Raven Press, New York, p 57

Correns C (1900) Gregor Mendels Regel ueber das Verhalten der Nachkommenschaft der Rassenbastarde. Ber dtsch bot Ges 18:158–168

Cuenot L (1902) La loi de Mendel et l'hérédité de la pigmentation chez les souris. Arch Zool Exp Gén 10:27–30

deVries H (1900) Das Spaltungsgesetz der Bastarde. Ber dtsch bot Ges 18:83–90

Dobzhansky T (1976) The myths of genetic predestination and of *tabula rasa*. Persp Biol Med 19:156–170

Hwang SJ, Beaty TH, McIntosh I, Hefferon T, Panny SR (1998) Association between homeobox-containing gene MSX1 and the occurrence of limb deficiency. Am J Med Genet 75:419–423

Johannsen W (1909) Elemente der exakten Erblichkeitslehre. Fischer, Jena

Kozlova SI, Prytkov AN (1986) Clinical genealogical method in diagnosis of hereditary diseases (Russian). In: Diagnosis of hereditary diseases. The Academy of Medical Sciences of the USSR, Moscow, pp 5–18

Mendel G (1866) Versuche ueber Pflanzenhybriden. Verh naturforsch Ver Bruenn 4:3–47

Mignot E (1998) Genetic and familial aspects of narcolepsy. Neurology 50 (2 Suppl 1): 16–22

Morgan TH, Sturtevant AH, Muller HJ, Bridges CB (1915) The mechanism of Mendelian heredity. H Holt & Co, New York

Plate L (1910) Vererbungslehre und Deszendenztheorie. Festschr f R Hertwig II Fischer, Jena, 537–610

Sutton WS (1903) The chromosomes in heredity. Biol Bull Wood's Hole 4:231–251

Timofeeff-Ressovsky NW (1925) On phaenotypic manifestation of genotype (Russian). J Exp Biol I: 93–142

Timofeeff-Ressovsky NW (1934) Verknuepfung von Gen- und Aussenmerkmal. Wiss Woche Frankfurt 1:92–115

Timofeeff-Ressovsky NW (1940) Allgemeine Erscheinungen der Gen-Manifestierung. Handbuch der Erbbiologie des Menschen Springer, Berlin, S 32–72

Timofeeff-Ressovsky NW, Vogt O (1926) Ueber idiosomatische Variationsgruppen und ihre Bedeutung fuer die Klassifikation der Krankheiten. Naturwiss 14:1188–1190

Vasku A, Soucek M, Znojil V, Rihacek I, Tschoplova S, Strelcova L, Cidl K, Blazkova M, Hajek D, Holla L, Vacha J (1998) Angiotensin I-converting enzyme and angiotensinogen gene interaction and prediction of essential hypertension. Kidney Int 53:1479–1482

von Tschermak E (1900) Ueber kuenstliche Kreuzung bei *Pisum sativum*. Ber dtsch bot Ges 18:232–239
Willing M, Sowers M, Aron D, Clark MK, Burns T, Bunten C, Crutchfield M, D'Agostino D, Jannausch
    M (1998) Bone mineral density and its change in white women: estrogen and vitamin D receptor
    genotypes and their interaction. J Bone Min Res 13:695–705

Genes Protecting Against Cancers and Tumor Suppressor Genes

J. J. Mulvihill

Summary

How many genes contribute to the susceptibility or resistance to neoplasia in human beings? The answer has implications for clinical cancer management, genetic counseling, and research into the origins and pathogenesis of specific cancers. One of the oldest catalogs of human disease genes, McKusick's *Mendelian Inheritance in Man (MIM)*, has an up-to-date on-line version, *OMIM* (www.ncbi.nlm.nih.gov/Omim). As of 15 February 2000, *OMIM* contained 11,201 entries, 11–14 % of the estimated number of human genes. By reading the hard copy volume of *MIM* and the periodic literature, as well as an extensive electronic search of *OMIM* and PUBMED from the US National Library of Medicine, it was concluded that, in 1999, 635 entries related to neoplasia (5.7 % of known human genes, 0.6 to 0.8 % of the total genome). About two-thirds of the traits are phenotypes, mostly clinical syndromes that predispose to or are complicated by malignant or benign neoplasia, such as neurofibromatosis, cystic fibrosis, and Cowden disease. These traits should be sought in a patient presenting which a specific tumor as a clue to etiology since they may represent germline mutations that have implications for genetic counseling of the patient's family. About one-third of the entries are protooncogenes, tumor suppressor genes, translocation breakpoints, fusion proteins, and other markers that have been seen to date only as somatic cell mutations, largely in sporadic tumors and cell lines. They may eventually be shown to have germline mutations but, in any case, contribute to pathogenesis of specific sporadic tumors. Since human cancer genes are recognized by the occurrence of disease when the normal allele is mutated, the photographic image is the maximal repertory of genes that contribute to resistance to cancer. Search of *OMIM* can yield a quick estimate of the upper limit of the number of genes contributing to specific tumors, for example breast cancer has 157 entries, colon cancer 118, pheochromocytoma 39, and pancreatic cancer 67.

Environmental Carcinogens

Despite the theme of this volume, we must start, in considering the resistance to cancer, with the well-known list of 27 substances, groups of substances, medical treatments or occupational exposures that are clearly carcinogenic in human

V. Boulyjenkov, K. Berg, Y. Christen (Eds.)
Genes and Resistance to Diseases
© Springer-Verlag Berlin Heidelberg 2000

beings, usually because they are mutagenic (Monson 1996; US DHHS 1998). And, another 130 entries may reasonably be anticipated to be carcinogenic according to the US National Toxicology Program (US DHHS 1998). These agents are strong human carcinogens that should be avoided to help us escape from getting cancer. There are surely anticarcinogens suspected of contributing to resistance to cancer and the anticarcinogenic drug tamoxifen gives proof of the principle that chemoprevention of cancer is possible.

In any case, even geneticists must continue to press for the control of environmental carcinogens, especially tobacco use in all its forms. Yet, everyone knows a 90-year-old Uncle Al who smoked cigarettes all his life and died after a broken hip. So, not every human being exposed to these carcinogens gets cancer.

Clinical Pharmacogenetics and Ecogenetics

Why? One explanation lies in the many metabolic steps between external exposure of the body to the occurrence of internal metastatic malignancy. One long established category, then, of genes for cancer resistance are the so-called pharmacogenetic traits (Evans 1993; Kalow 1992; Weber 1997). Often these traits are xenobiotic metabolizing enzymes that are polymorphic and have one allele that confers relative resistance to cancer. Obviously, it is usually stated that one allele predisposes to cancer. One could use the analogy that the list of predisposing alleles is a photographic negative and that the positive print is the list of the normal alleles that confer resistance to cancer. Such polymorphic differences are associated with moderately increased relative risks in case-control studies. Still, in considering total populations, the impact of such cancer resistance alleles is large, since the resistance alleles may be quite frequent.

In addition to the pharmocogenetic resistance to carcinogens, there is a short list of clinical traits that enormously predispose to cancer after exposure to environmental agents. Consider skin cancer and ultraviolet light exposure in oculocutaneous albinism. In contrast to light-skinned Caucasians, Africans (Negros) have a normal skin color that is black or dark brown and which protects against the mutagenicity and carcinogenicity of sunlight. Whites lack the pigmentary granules in melanocytes that protect the skin cell nuclei of blacks. Some 22 clinical genetic traits predispose to cancer due to specific agents, whether physical, chemical, dietary, or microbial (Table 1). The wild type of those ecogenetic traits, therefore, confers resistance to cancer.

Four clinical genetic disorders have been suspected of having a deficit of cancer for unclear reasons: osteogenesis imperfecta, schizophrenia, Huntington disease, and Down syndrome. One of the leaders in familial cancer research, Henry Lynch (1966), published on families with osteogenesis imperfecta, brittle bone disease. He reported a nuclear family with osteogenesis imperfecta with no cases of cancer, which had been common in the older generation with osteogenesis imperfecta (Lynch et al. 1966). The suggestion is quite old and not addressed by recent epidemiologic methods. Still, with current genetic epidemiologic tech-

Table 1. Clinical Ecogenetics of Human Tumors

Environmental agent	Genetic trait	Tumor/outcome
Radiation, ultraviolet	Oculocutaneous albinism Xeroderma pigmentosum Dysplastic nevus syndrome	Skin cancer Skin cancer Melanoma
Radiation, ionizing	Nevoid basal cell carcinoma syndrome Retinoblastoma Li-Fraumeni (SBLA) cancer family syndrome Ataxia-telangiectasia with lymphoma	Basal cell carcinoma Sarcoma Sarcoma, breast cancer Radiation toxicity
Androgen	Fanconi pancytopenia	Hepatoma
Dexamethasone	Dexamethasone-sensitive aldosteronism	Adrenal cortical adenoma
Stilbestrol	Turner syndrome of gonadal dysgenesis	Adenosquamous endometrial carcinoma
Iron	Hemochromatosis	Hepatocellular carcinoma
Tyrosine	Tyrosinemia	Hepatocellular carcinoma
Diet	Polyposes coli Barrett esophagus Lewis antigen (a–b–)(fucosyltransferase)	Colonic and gastric carcinomas Esophageal adenocarcinoma Alimentary tract carcinomas
Epstein-Barr Virus	X-linked lymphoproliferative syndrome HLA-A2; Bw46	Burkitt and other lymphomas Nasopharyngeal carcinoma
Papilloma virus type 5	Epidermodysplasia verruciformis	Skin cancer
Hepatitis B virus	Virus integration site	Hepatocellular carcinoma
N-substituted aryl compounds	N-acetyltransferase activity (NAT2)	Urinary bladder carcinoma
Tobacco smoke	Cytochrome P450dbl (CYP2D6) Cytochrome P_l450 (CYP1A1)	Lung cancer Lung, laryngeal, oral, breast, and urinary bladder cancers

niques and population-based cancer incidence rates, the hypothesis could be addressed with vigor on whether any well registered mendelian trait has a deficit (or excess) of cancer in general or, if numbers permit, of any particular tumor type.

Trinucleotide Repeats and Cancer Resistance

A more intriguing hypothesis with data that are much better, but still controversial, is the possible deficit of cancer in patients with schizophrenia (Gulbinat et al. 1992; Mortensen 1994; Saku et al. 1995). Sørensen with others in Copenhagen suggested a deficit of deaths from cancer among Huntington disease patients (Sørensen and Fenger 1992). In such severe neuropsychiatric disorders as schizophrenia and Huntington disease, there is a real possibility of a bias from failing to recognize cancer in hopelessly ill individuals. But the excellent databases of the Danish Cancer Registry and the Danish Registry of Huntington Disease show a deficit of cancer based on both death certificates and cancer incidence among persons with Huntington disease. Recently, Sørensen et al. (1999) showed that cancer incidence is only 60 % of normal expectation.

It is provocative that the trinucleotide repeat expansion that produces the abnormal huntingtin protein and is associated with apoptosis seems to offer protection from cancer. This speculation is all the more intriguing because another gene, the androgen receptor gene, *AR*, has a major role in prostate cancer (as discussed below). This gene is constitutionally mutated in testicular feminization patients and in those with Kennedy type spinal and bulbar atrophy. The neurologic syndrome has not been associated with cancer, but perhaps the question should be addressed formally. Female-appearing XY individuals with testicular feminization have been reported to have breast cancer (Wooster et al. 1992); but, rates are not known. So, the occurrence of breast cancer may be an excess for men, which such individuals are genetically, or a deficit for women, which they are by appearance.

More intriguing, however, is the correlation between the number of CAG repeats in the germline *AR* and the risk of prostrate cancer (Cude 1999; Edwards et al. 1999). Specifically, a greater number of repeats is associated with a lower risk or less virulent prostate cancer. Further, in population studies, African-Americans have a higher risk of prostate cancer than men of other races and, as a group, a lower number of repeats in *AR*. So, long repeats of CAG in *AR* contribute to resistance to prostate cancer. These observations could be extended by epidemiological studies of cancer in other trinucleotide repeat syndromes (Wells and Warren 1998) or with in vitro studies of such expansions in cancer cells.

Down Syndrome

Early case reports and case series suggested a considerate risk for leukemia in children with Down syndrome, due to a third copy of at least part of chromosome 21 (Krivit and Good 1957; Miller 1963). Recently, the excellent record linkage systems for epidemiologic research in Denmark have quantified the risk and extended it to include the growing population of adults with Down syndrome (Hasle 2000). Matching the Cytogenetic Register, the Cancer Registry, and the Central Population Register in Denmark, they identified 60 Down syndrome individuals with malignancy, when 49.8 were expected, giving a standardized incidence ratio (SIR) of 1.2 (not statistically different from 1.0). Significant differences were seen as large excesses of leukemia (SIR of 56 for ages 0 to 4 years; 10 for ages 5 to 29 years). Under 5 years, the SIR was 41 for acute lymphocytic leukemia and 154 for acute myelogenous leukemia. Hence, the risk for leukemia is confined to childhood and adolescent, with no case occurring after age 29 years.

The startling finding, yet to be repeated on such a scale, was the occurrence of only 24 solid tumors despite an expectancy of 47.8. Whereas 7.3 cases of breast cancers were expected, none was seen. Other voids were for oral cancer and Hodgkin's and non-Hodgkin lymphomas. The overall deficit of solid tumors was statistically significant past age 30. Since most cancers in adult life are carcinomas, the authors raised the possibility that trisomy 21 has a gene dosage effect of a tumor suppressor gene for epithelial tumors on chromosome 21. They even offer the candidate of the copper-zinc superoxide dismutase.

In short, there are skimpy details about the clinical genetics of cancer resistance. Just as there are cancer-prone families, there must surely be cancer-free families. But, some of each type must occur by chance. And, while it has been arduous but productive to define and delineate true cancer family syndromes, the notion of cancer-resistant families is saved for speculation at research workshops on cancer genetics.

Catalog of Human Cancer Genes

To be systematic about thinking of genes that protect against cancer, including the clearly named tumor suppressor genes, a geneticist turns first to that bible of medical genetics, McKusick's *Mendelian Inheritance in Man (MIM)* (McKusick 1998), or, more specifically, its online version, *OMIM*. As of 15 February 2000, there were 11,201 entries in *OMIM* (OMIM). Since there are an estimated 80,000 to 100,000 human genes, one could say that some 11 to 14 % of functional genes have been identified, at least in that database, which includes clinical phenotypes and clearly functional genes, as opposed to other databases, which include just RNAs or expressed sequence tags (ESTs).

For 25 years, I have tried to keep up with the entries in *MIM* and *OMIM* that relate to cancer and neoplasin (Table 2). Over those decades, the total number of entries has grown exponentially. The percentage of the total number of genes associated with cancer has fallen slightly from 8.6 % to 6.2 %, largely because of the very high rate of discovery of genes or proteins with no assigned phenotype. Looked at in another way, making the denominator the estimated total number of human genes, then some 0.6 to 0.8 % of human genes, when mutated, have been associated with neoplasms. I have recently published a catalog of these human

Table 2. Catalog of Human Cancer Genes, 1971 to 1999

McKusick Edition-year	Mulvihill year	Number of entries		% with Neoplasia
		Neoplasia	Total	
3rd-1971	1975	161	1876	8.6
4th-1975	1977	200	2336	8.6
8th-1988	1989	339	4345	7.8
11th-1994	1995	467	6678	7.0
12th-1999	1999	635	10146	6.2

Table 3. *Catalog of Human Cancer Genes* (Mulvihill, 1999)

Category	Number	Category	Number
Phakomatoses	30	Protooncogenes	54
Nervous system	26	Chromosomal breakpoints	63
Endoerine System	43	Tumor suppressor genes	28
Mesodermal tissue	29	Proteins, other markers	**68**
Alimentary system	48	Subtotal (Gene markers)	213
Urogenital system	45		
Cardiorespiratory system	16		
Skeletal system	35		
Cutaneous system	79		
Hemolymphatic system	71		
Subtotal (Phenotypes)	422	Grand Total	635

traits and genes, annotating each entry with my personal experience (Mulvihill 1999). The *Catalog* has 635 traits and genes associated with neoplasia (Table 3).

This number was derived by reading *MIM*, electronically searching *OMIM* and keeping up with the periodic literature. The figure includes more than frank malignancy; it includes tumors, both malignant or benign, that are the sole feature of an *OMIM* entry, a frequent concomitant, or just a rare complication, sometimes reported as a single case report only. To the extent that my volume includes human genes that mostly have disease phenotypes, namely neoplasia, when they are mutated, it is the photographic negative of genes that contribute to resistance to neoplasia. Since 90 % of human genes are yet to be identified or associated with cancer, there could obviously be a large number of cancer resistance genes. With the extreme case that all cancer-predisposing genes have been identified, over 99 % of the genes may help resist cancer! It is clearly plausible that many genes do not modify cancer risk at all. In either extreme, we have just begun identifying cancer resistance genes.

Surfing *OMIM* reveals a large list of genes that are mutated almost exclusively in the somatic cell of cancer and not at the germ cell level. In my *Catalog* (Mulvihill 1999), such genes are mostly grouped in a final chapter because, if they were seen as germline mutations, they usually have a specific phenotype that can be assigned to an organ system chapter. So minimal numbers of genes seen to date chiefly as somatic cell mutations include 54 protooncogenes, 28 tumor suppressor genes, 63 translocation break points, and 68 protein and other markers. The translocation break points and other cytogenetic changes associated with human neoplasia have been catalogued elsewhere (Mitelman et al. 1997).

In 1914, Boveri suggested that all cancers have abnormal chromosomes. The recent fusion of classical cytogenetics with molecular genetics has given insight to the fusion proteins that are generated by the breaking of two chromosomes and the exchange and re-annealing of the broken ends. The classic illustration is the Philadelphia chromosome, the 9;22 translocation seen in nearly all cases of chronic myelogenous leukemia. The fusion proteins and the chromosomal breakpoints have been associated mostly with leukemias, because these malignancies are easier to biopsy, purify, and analyze than solid tumors. Nonetheless, it is very intriguing that even benign neoplasms, for example, salivary gland tumors, lipomas, and leiomyomas, also have translocation break points that are leading to the identification of genes related to the development of benign tumors. How translocating chromosomes relates to genes for resistance to cancer is either very obvious or very obscure. That is to say, the creation of such fusion proteins clearly initiates or promotes the carcinogenic process; hence, any insight into the normal proteins from which they are made would be an insight into the mechanisms of resistance to cancer development.

Tumor Suppressor Genes

I have emphasized areas of tumor resistance genes other than the frank tumor suppressor genes, because I think every scientist interested in genetics or cancer knows something about tumor suppressor genes. They were first defined by Harris et al. (1969), who studied artificially fused cells and found combinations of hybrid cells that suppressed the in vitro phenotype of cancer. This laboratory phenotype gained theoretical underpinnings with Knudson's now legendary hypothesis of the two hits origin of human cancers starting with the simple case of retinoblastoma (1971). The functional definition of a tumor suppressor gene now is the loss of genetic heterogeneity, that is the demonstration of only one active allele in tumor cells from a person known to have two active alleles in normal tissue.

Perhaps the most comprehensively studied model of a tumor suppressor gene is *TP53*, and justly so, although it also has characteristics of a protooncogene. It is somatically abnormal in many human cancers and constitutionally mutated in the Li-Fraumeni cancer family syndrome or the SBLA syndrome. The acronym is an aide to remember the tumor types: Sarcoma, Breast, Bone, and Brain tumors, Leukemia, Laryngeal, and Lung cancers, and Adrenal cortical neoplasms. The p53 protein has been called the guardian of the genome and is a transcription factor that, when present in normal form and amount, arrests the cell cycle as if to allow DNA damage to be repaired or apoptosis to begin. The defining features of a tumor suppressor gene are phenomenologic, namely the demonstration of loss of genetic heterogeneity and the experimental demonstration that reintroduction of the normal gene suppresses the malignant phenotype in vitro.

As an illustration of the search power of *OMIM*, one can electronically scan in milliseconds the database that is now 3 printed volumes, for terms such as "loss of heterogeneity", "LOH", and delet* (the asterisk allows various endings of that root) or "suppress*". By these strategies, some 229 entries were in *OMIM* as of November 4, 1999, including some very well known ones and many remaining to be classified. All of these deserve exploration by available techniques. Some may explain aggregations of familial cancer not yet documented or even clearly defined in clinical entities. Any of them could be a way to define the normal allele that has been contributing to cancer resistance.

Summary Comment

The classic paradigm of carcinogenesis speaks of initiation and promotion and metastasis. In reviewing cancer resistance genes, I am intrigued by a contrast or paradox. Much research has focused on the beginnings of cancer to understand the process that is initiation, that is the point where an environmental factor interacts with a cell and its genome to cause a mutation that begins a normal cell down the pathway to an abnormal cancer cell and a metastatic lethal mass. We read from left to right in most of the western world. We think that understanding mutation and cancer initiation is the preferred route to understanding and eventually controlling human cancer.

But, many people read from right to left, especially many cancer epidemiologists, even in the West, who think the way to decrease the human burden of cancer is to focus on issues of the late stages of carcinogenesis, mainly metastasis and late growth. Hence, we saw in 1998 the public infatuation with angiogenesis and antiangiogenesis as ways to control clinical cancers. After all, what is the harm if we harbor a few clusters of cancer cells in our body? They are no problem, if they would only not progress and metastasize.

So really the question I come to is this: what genes can lead to the resistance to progression and metastasis? *TP53* if often mutated late in colon carcinogenesis and perhaps others. As shown by microarray chip technology, a large number of genes are differentially expressed in metastatic cancer but not in the early cancer. One goal of the US National Cancer Institute's Genetic Anatomy Project (www.ncbi.nlm.nih.gov/CGAP) is to show these differences. There are predictions that such analyses will be as standard as histopathologic descriptions. The expectation is that the expressed genetic signature of every individual's peculiar cancer will give a clue to etiology and will guide individualized therapies, including designer molecules that are tailored to the genetic pattern of each person's own cancer. Then, the interface between in vivo and in vitro cancer predisposition and cancer resistance will truly vanish, as may the burden of human cancer.

Acknowledgment

I thank Ashley Weedn for manuscript preparation. Supported in part with U.S. National Cancer Institute grant CA 75311.

References

Cude KJ, Dixon SC, Guo Y, Lisella J, Figg WD (1999) The androgen receptor: genetic considerations in the development and treatment of prostate cancer. J Mol Med 77:419–426

Edwards SM, Badzioch MD, Minter R, Hamoudi R, Collins N, Ardern-Jones A, Dowe A, Osborne S, Kelly J, Shearer R, Easton DF, Saunders GF, Dearnaley DP, Eeles RA (1999) Androgen receptor polymorphisms: association with prostate cancer risk, relapse and overall survival. Int J Cancer 84:458–465

Evans DA (1993) Genetic factors in drug therapy: clinical and molecular pharmacogenetics. Cambridge University Press, Cambridge

Gulbinat W, Dupont A, Jablensky A, Jensen OM, Marsella A, Nakane Y, Sartorius N (1992) Cancer incidence of schizophrenic patients. Results of record linkage studies in three countries. Br J Psychiatry Suppl 18:75–83

Harris H, Miller OJ, Klein G, Worst P, Tachibana T (1969) Suppression of malignancy by cell fusion. Nature 223:363–368

Hasle H, Clemmensen IH, Mikkelsen (2000) Risks of leukaemia and solid tumours in individuals with Down's syndrome. Lancet 355:165–169

International Agency for Research on Cancer (1998) Overall evaluations of carcingenicity to humans – IARC monographs programme on the evaluation of carcinogenic risks to humans. IARC, Lyon, France Retrieved from the World Wide Web: http://www.iarc.fr/

Kalow W (1992) (ed) Pharmacogenetics of drug metabolism. Pergamon Press, New York

Knudson AG (1971) Mutation and cancer: statistical study of retinoblastoma. Proc Natl Acad Sci 68:820–823

Krivit W, Good RA (1957) Simultaneous occurrence of mongolism and leukemia. Am J Dis Child 94:289–298

Lynch HT, Lemon HM, Krush AJ (1966) A note on "cancer-susceptible" and "cancer-resistant" genotypes: implications for cancer detection and research. Nebr State Med J 51:209–211

McKusick VA (1962) On the X chromosome of man. Quart Rev Biol 37:69–175

McKusick VA (1966) Mendelian inheritance in man: catologs of autosomal dominant, autosomal recessive, and x-linked phenotypes. 1st Ed. John Hopkins Press, Baltimore

McKusick VA (1971) Mendelian inheritance in man: catologs of autosomal dominant, autosomal recessive, and x-linked phenotypes. 3rd Ed. John Hopkins Press, Baltimore, ix

McKusick VA (1974) Mendelian inheritance in man: catologs of autosomal dominant, autosomal recessive, and x-linked phenotypes. 4th Ed. John Hopkins University Press, Baltimore, xii

McKusick VA (1988) Mendelian inheritance in man: catologs of autosomal dominant, autosomal recessive, and x-linked phenotypes. 8th Ed. John Hopkins University Press, Baltimore, xi

McKusick VA (1994) Mendelian inheritance in man: catologs of autosomal dominant, autosomal recessive, and x-linked phenotypes. 11th Ed. John Hopkins University Press, Baltimore, vii

McKusick VA (1998) Mendelian inheritance in man: catologs of autosomal dominant, autosomal recessive, and x-linked phenotypes. 12th Ed. John Hopkins University Press, Baltimore

Miller RW (1963) Down's syndrome (mongolism), other congenital malformations and cancers among the sibs of leukemic children. N Engl J Med 268:393–401

Mitelman F, Mertens F, Johansson B (1997) A breakpoint map of recurrent chromosomal rearrangements in human neoplasia. Nat Genet 15:417–474

Monson RR (1996) Occupation. In: Schottenfeld D, Fraumeni JF Jr (eds) Cancer epidemiology and prevention. Oxford University Press, New York, 373–405

Mortensen PB (1994) The occurrence of cancer in first admitted schizophrenic patients. Schizophr Res 12:185–194

Mulvihill JJ (1975) Congenital and genetic diseases. In Fraumeni JF Jr (ed) Persons at high risk of cancer: an approach to cancer etiology and control. Academic Press, New York, 3–37

Mulvihill JJ (1977) Genetic repertory of human neoplasia. In Mulvihill JJ, Miller RW, Fraumeni JF Jr (eds) Genetics of human cancer. Raven Press, New York, 137–143

Mulvihill JJ, Miller RW, Fraumeni JF Jr (1977) (eds) Genetics of human cancer. Raven Press, New York

Mulvihill JJ (1989) Prospects for cancer control and prevention through genetics. Clin Genet 36:313–319

Mulvihill JJ (1994) Clinical ecogenetics of human cancer. Hem/Onc Ann 2:157–161

Mulvihill JJ, Davis S, Fromkin KR (1996) The catalog of human genes predisposing to neoplasia. In: Weber W, Narod S, Mulvihill JJ (eds) Familial cancer management. CRC Press, Boca Raton, 203–237

Mulvihill JJ (1999) Catalog of human cancer genes. Johns Hopkins University Press, Baltimore

Saku M, Tokudome S, Ikeda M, Kono S, Makimoto K, Uchimura H, Mukai A, Yoshimura T (1995) Mortality in psychiatric patients, with a specific focus on cancer mortality associated with schizophrenia. Int J Epidemiol 24:366–372

Sørensen SA, Fenger K (1992) Causes of death in patients with Huntington's disease and in unaffected first degree relatives. J Med Genet 29:911–914

Sørensen SA, Fenger K, Olsen JH (1999) Significantly lower incidence of cancer among patients with Huntington disease: an apoptotic effect of an expanded polyglutamine tract? Cancer 86:1342–1346

United States Department of Health and Human Services, Public Health Service, National Toxicology Program (1998) Eighth report on carcinogens-internet summary. National Toxicology Program, Research Triangle Park, N.C. Retrieved from the World Wide Web: http://ehis.niehs.nih.gov/roc/

Weber W (1997) (ed) Pharmacogenetics. Oxford University Press, New York

Wells RD, Warren ST (1998) (eds) Genetic instabilities and hereditary neurological diseases. Academic Press, San Diego

Wooster R, Mangion J, Eeles R, Smith S, Dowsett M, Averill D, Barrett-Lee P, Easton DF, Ponder BAJ, Stratton MR (1992) A germline mutation in the androgen receptor gene in two brothers with breast cancer and Reifenstein syndrome. Nat Genet 2:132–134

Subject Index

ABCR 21
adeno-associated viral vector 153–154
adenoviral vector 150–152
age-related macular degeneration 19–26
aging 1–7, 19, 20, 22, 122, 128
Aids 9–17, 148
Alzheimer's disease 20, 24, 121–131, 138
amyloid 121–129
amyotrophic lateral sclerosis 123
angiotensin I converting enzyme 84, 86, 163
antagonistic pleiotropy 4, 5
apolipoprotein A-I 51–54, 67–81
apolipoprotein B 53, 83, 84, 163
apolipoprotein E 19, 20, 22–25, 72, 84, 121, 126
apoptosis 94, 95, 98
atherosclerosis 51–65, 67–81

cancer 27–42, 138, 148, 164, 169–178
carcinogenesis 27–42, 169, 170
cardiovascular disease 51–90, 163, 164
CCR5 9, 11–14
chemokine 12, 13, 94
cholesterol Ester Transfer Protein 51–65, 84, 86
cholesterol transport 52, 53
compensatory adaptation theory of aging 2–5
Creutzfeldt-Jakob disease 123
cytochrome P450 28–38, 171

depression 136–138
DNA repair gene 4
Down syndrome 170, 172, 173

ecogenetics 170, 171
Ehlers-Danlos syndrome 162, 163
endothelial cell 91–103

free radicals 21, 127
frontotemporal dementia with parkinsonism 123, 127

gene affecting cognitive and emotional functions 133–145
gene involved in resistance to carcinogenesis 27–42
gene protecting against age-related macular degeneration 19–26
gene protecting against cancers 169–178
gene that limit aids 9–17
gene therapy 147–157
genetic factors in malaria resistance 105–119
genetic of longevity 1–7, 121
glucose 6-phosphate dehydrogenase 108, 110–116
glutathione S-transferase 36–40
good gene/bad gene 87, 88, 159–168

hemoglobin S 110–112
hemo oxygenase-1 95–98
hemophilia 153, 154
high density lipoprotein (HDL) 51–57, 59, 60, 62, 67–78, 87
HIV 9–17, 149, 150
HLA 9–11, 14, 140, 165
human genome 10, 88
Huntington's disease 123, 170–172

individuality in response to toxins and
  carcinogens 28, 30
insulinlike growth factor-2 receptor (and
  "gene for" IQ) 141
IQ 134–136, 141, 142

lentiviral vector 149–151
leptin 43–49
Limone sul Garda 68–70, 78
linkage disequilibrium 133, 140
low-density lipoprotein (LDL) 22–25, 52,
  55, 56, 60
Lp(a) lipoprotein 51, 58

malaria resistance 105–119
myocardial infarction 57, 58, 61

neural circuit regulating body weight
  43–49
neurodegenerative disease 20, 24,
  121–131
NF-ϰB 94, 95, 97

obesity 43–49

Parkinson's disease 123
pharmacogenetic 35, 128, 170, 171
presenilin 121–126

protective response of endothelial cell
  91–103

quantitative trait loci 133, 136, 139, 141,
  142

rate-of-living theory 5, 6
response modifier 85, 86
retroviral vector 148–149

schizophrenia 170, 171
selection in human evolution 105–108,
  128
serotonin transporter gene 140, 141

tau 121–123, 127, 128
α-thalassemia 109, 110
toxin 28, 30
trinucleotide repeats 171, 172
tumor suppressor gene 169–178
twin 57, 134–137

variability gene concept 55–57
variation of hereditary traits 160–163
very low density lipoprotein (VLDL) 52

xenograft rejection 91–99

Printing: Saladruck, Berlin
Binding: H. Stürtz AG, Würzburg